THE MEANING *of* HUMAN EXISTENCE

ALSO BY EDWARD O. WILSON

A Window on Eternity: Gorongosa National Park, Mozambique (2014)

Letters to a Young Scientist (2013)

Why We Are Here: Mobile and the Spirit of a Southern City,
with Alex Harris (2012)

The Social Conquest of Earth (2012)

Kingdom of Ants: José Celestino Mutis and the Dawn of Natural History in the New World, with José M. Gómez Durán (2011)

The Leafcutter Ants: Civilization by Instinct, with Bert Hölldobler (2011)

Anthill: A Novel (2010)

The Superorganism: The Beauty, Elegance and Strangeness of Insect Societies,
with Bert Hölldobler (2009)

The Creation: An Appeal to Save Life on Earth (2006)

Nature Revealed: Selected Writings, 1949–2006 (2006)

From So Simple a Beginning: The Four Great Books of Darwin,
edited with introductions (2005)

Pheidole in the New World: A Dominant, Hyperdiverse Ant Genus (2003)

The Future of Life (2002)

Biological Diversity: The Oldest Human Heritage (1999)

Consilience: The Unity of Knowledge (1998)

In Search of Nature (1996)

Naturalist (1994); new edition, 2006

Journey to the Ants: A Story of Scientific Exploration,
with Bert Hölldobler (1994)

The Diversity of Life (1992)

Success and Dominance in Ecosystems: The Case of the Social Insects (1990)

The Ants, with Bert Hölldobler (1990); Pulitzer Prize, General
Nonfiction, 1991

Biophilia (1984)

Promethean Fire: Reflections on the Origin of Mind,
with Charles J. Lumsden (1983)

Genes, Mind, and Culture: The Coevolutionary Process, with Charles J.
Lumsden (1981)

Caste and Ecology in the Social Insects, with George F. Oster (1978)

On Human Nature (1978); Pulitzer Prize, General Nonfiction, 1979

Sociobiology: The New Synthesis (1975); new edition, 2000

The Insect Societies (1971)

A Primer of Population Biology, with William H. Bossert (1971)

The Theory of Island Biogeography, with Robert H. MacArthur (1967)

Edward O. Wilson

THE MEANING *of* HUMAN EXISTENCE

LIVERIGHT PUBLISHING CORPORATION

A Division of W. W. Norton & Company   New York · London

For information about permission to reproduce selections from this book,
write to Permissions, Liveright Publishing Corporation,
a division of W. W. Norton & Company, Inc.,
500 Fifth Avenue, New York, NY 10110

For information about special discounts for bulk purchases,
please contact W. W. Norton Special Sales
at specialsales@wwnorton.com or 800-233-4830

Manufacturing by Courier Westford
Book design by Chris Welch Design
Production manager: Devon Zahn

Library of Congress Cataloging-in-Publication Data

Wilson, Edward O., author.
The meaning of human existence / Edward O. Wilson. — First edition.
    pages cm
ISBN 978-0-87140-100-7 (hardcover)
1. Philosophical anthropology. I. Title.
BD450.W5225 2014
128—dc23

                                                      2014016707

Liveright Publishing Corporation
500 Fifth Avenue, New York, N.Y. 10110
www.wwnorton.com

W. W. Norton & Company Ltd.
Castle House, 75/76 Wells Street,
London W1T 3QT

1 2 3 4 5 6 7 8 9 0

Contents

I

THE REASON WE EXIST

HISTORY MAKES LITTLE SENSE WITHOUT
PREHISTORY, AND PREHISTORY MAKES LITTLE
SENSE WITHOUT BIOLOGY. KNOWLEDGE OF
PREHISTORY AND BIOLOGY IS INCREASING
RAPIDLY, BRINGING INTO FOCUS HOW HUMANITY
ORIGINATED AND WHY A SPECIES LIKE OUR OWN
EXISTS ON THIS PLANET.

The Meaning of Meaning

Does humanity have a special place in the Universe? What is the meaning of our personal lives? I believe that we've learned enough about the Universe and ourselves to ask these questions in an answerable, testable form. With our own eyes we can see through the dark glass, fulfilling Paul's prophecy, "Now I know in part; then I shall know fully, even as I am fully known." Our place and meaning, however, are not being revealed as Paul expected—not at all. Let's talk about that, let us reason together.

I propose a journey to this end, in which I ask to serve as guide. Our route will pass first through the origin of our species and its place in the living world, questions I initially addressed in a different context in *The Social Conquest of Earth*. Then it will approach, through selected steps from the natural sciences into the humanities and back again, the more difficult problem of "Where are we going?" and the most difficult question of all, "Why?"

The time has come, I believe, to make a proposal about the possibility of unification of the two great branches of learning. Would the humanities care to colonize the sciences? Maybe use a little help doing that? How about replacing science fiction, the imagining of fantasy by a single mind, with new worlds of far greater diversity based on real science from many minds? Might poets and visual artists consider searching in the real world outside the range of ordinary dreams for unexplored dimensions, depth, and meaning? Would they be interested in finding the truth of what Nietzsche called, in *Human, All Too Human*, the rainbow colors around the outer edges of knowledge and imagination? That is where meaning is to be found.

In ordinary usage the word "meaning" implies intention, intention implies design, and design implies a designer. Any entity, any process, or definition of any word itself is put into play as a result of an intended consequence in the mind of the designer. This is the heart of the philosophical worldview of organized religions, and in particular their creation stories. Humanity, it assumes, exists for a purpose. Individuals have a purpose in being on Earth. Both humanity and individuals have meaning.

There is a second, broader way the word "meaning" is used and a very different worldview implied. It is that the accidents of history, not the intentions of a designer, are the source of meaning. There is no advance design, but instead overlapping networks of physical cause and effect. The unfolding of history is obedient only to the general laws of the Universe. Each event is random yet alters the probability of later events. During organic evolution, for example, the origin of one adaptation by natural selection makes the origin of certain other adaptations more likely. This concept of meaning, insofar as it illuminates humanity and the rest of life, is the worldview of science.

Whether in the cosmos or in the human condition, the second, more inclusive meaning exists in the evolution of present-day reality amid countless other possible realities. As more complex biological entities and processes arose in past ages, organisms drew closer in their behavior to include the use of intentional meaning: at first there were the sensory and nervous systems of the earliest multicellular organisms, then an organizing brain, and finally behavior that fulfills intention. A spider spinning its web intends, whether conscious of the outcome or not, to catch a fly. That is the meaning of the web. The human brain evolved under the same regimen as the spider's web. Every decision made by a human being has meaning in the first, intentional sense. But the capacity to decide, and how and why the capacity came into being, and the consequences that followed, are the broader, science-based meaning of human existence.

Premier among the consequences is the capacity to imagine possible futures, and to plan and choose among them. How wisely we use this uniquely human ability depends on the accuracy of our self-understanding. The question of greatest relevant interest is how and why we are the way we are and, from that, the meaning of our many competing visions of the future.

The advances of science and technology will bring us to the greatest moral dilemma since God stayed the hand of Abraham: how much to retrofit the human genotype. Shall it be a lot, a little bit, or none at all? The choice will be forced on us because our species has begun to cross what is the most important yet still least examined threshold in the technoscientific era. We are about to abandon natural selection, the process that created us, in order to direct our own evolution by volitional selection— the process of redesigning our biology and human nature as we wish them to be. No longer will the prevalence of some genes (more precisely alleles, variations in codes of the same gene) over others be the result of environmental forces, most of which are beyond human control or even understanding. The genes and their prescribed traits can be what we choose. So—how about longer lives, enlarged memory, better vision, less aggressive behavior, superior athletic ability, pleasing body odor? The shopping list is endless.

In biology, how-and-why explanations are routine and expressed as "proximate" and "ultimate" causation of living processes. An example of the proximate is this: we have two hands and ten fingers, with which we do thus and so. The ultimate explanation is *why* we have two hands and ten fingers to start with, and why are we prone with them to do thus and so and not something else. The proximate explanation recognizes that anatomy and emotions are hardwired to engage in certain activities. The ultimate explanation answers the question, why this particular hardwiring and not some other? To explain the human condition, thereby to give meaning to the human existence, requires both levels of explanation.

In the essays to follow, I've addressed the second, broader meaning of our species. Humanity, I argue, arose entirely on its own through an accumulated series of events during evolution. We are not predestined to reach any goal, nor are we answerable to any power but our own. Only wisdom based on self-understanding, not piety, will save us. There will be no redemption or second chance vouchsafed to us from above. We have only this one planet to inhabit and this one meaning to unfold. To take this step in our journey, to get hold of the human condition, we need next a much broader definition of history than is conventionally used.

2

Solving the Riddle of
the Human Species

To grasp the present human condition it is necessary to add the biological evolution of a species and the circumstances that led to its prehistory. This task of understanding humanity is too important and too daunting to leave exclusively to the humanities. Their many branches, from philosophy to law to history and the creative arts, have described the particularities of human nature back and forth in endless permutations, albeit laced with genius and in exquisite detail. But they have not explained why we possess our special nature and not some other, out of a vast number of conceivable natures. In that sense, the humanities have not achieved nor will they ever achieve a full understanding of the meaning of our species' existence.

So, as best we can answer, just what are we? The key to the great riddle lies in the circumstance and process that created our species. The human condition is a product of history—not just the six millennia of civilization but very much further back, across hundreds of millennia. The whole of it, biological and cultural evolution, must be explored in seamless unity for a complete answer to the mystery. When viewed across its entire traverse, the history of humanity also becomes the key to learning how and why our species arose and survived.

A majority of people prefer to interpret history as the unfolding of a supernatural design, to whose author we owe obeisance. But that comforting interpretation has grown less supportable as knowledge of the real world has expanded. Scientific knowledge in particular, measured by numbers of scientists and scientific journals, has been doubling every ten to twenty years for over a century. In traditional explanations of the past, religious creation stories have been blended with the humanities to attribute meaning to our species' existence. The time has come to consider what science might give to the humanities and the humanities to science in a common search for a more solidly grounded answer than before to the great riddle of our existence.

To begin, biologists have found that the biological origin of advanced social behavior in humans was similar to that occurring elsewhere in the animal kingdom. Using comparative studies of thousands of animal species, from insects to mammals, we've concluded that the most complex societies have arisen through eusociality—meaning, roughly, the "true" social condition. By definition, the members of a eusocial group cooperatively rear the young across multiple generations. They also divide labor through the surrender by some members of at least part of their personal reproduction in a way that increases the "reproductive success" (lifetime reproduction) of other members.

Eusociality stands out as an oddity in a couple of ways. One is its extreme rarity. Out of hundreds of thousands of evolving lines of animals on the land during the past four hundred million years, the condition, so far as we can determine, has arisen only nineteen times, scattered across insects, marine crustaceans, and subterranean rodents. The number is twenty, if we include human beings. This is likely to be an underestimate, perhaps a gross one, due to sampling error. Nevertheless, we can be certain that the number of originations of eusociality was relatively very small.

Furthermore, the known eusocial species arose very late in the history of life. It appears to have occurred not at all during the great Paleozoic diversification of insects, 350 to 250 million years before the present, during which the variety of insects approached that of today. Nor is there as yet any evidence of eusocial species alive during the Mesozoic Era until the appearance of the earliest termites and ants between 200 and 150 million years ago. Humans at the Homo level appeared only very recently, following tens of millions of years of evolution among the Old World primates.

Once attained, advanced social behavior at the eusocial grade found a major ecological success. Of the nineteen known independent lines among animals, just two within the insects—ants and termites—globally dominate invertebrates on the land. Although they are represented by fewer than twenty thousand of the million known living insect species, ants and termites compose more than half of the world's insect body weight.

The history of eusociality raises a question: Given the enormous advantage it confers, why has this advanced form of social behavior been so rare and long in coming? The answer appears to be the special sequence of preliminary evolutionary changes that must occur before the final step to eusociality can be taken. In all of the eusocial species analyzed to date, the final step before eusociality is the construction of a protected nest, from which foraging trips are launched and within which the young are raised to maturity. The original nest builders can be a lone female, a mated pair, or a small and weakly organized group. When this final preliminary step is attained, all that is needed to create a eusocial colony is for the parents and offspring to stay at the nest and cooperate in raising additional generations of young. Such primitive assemblages then divide easily into risk-prone foragers and risk-averse parents and nurses.

What brought a single primate line to the rare level of eusociality? Paleontologists have found that the circumstances were humble. In Africa roughly two million years ago, one species of the primarily vegetarian australopithecines evidently began to shift its diet to include a much higher reliance on meat. For a group to harvest such a high-energy, widely dispersed source of food, it did not pay to roam about as a loosely organized pack of adults and young in the manner of present-day chimpanzees and bonobos. It was more efficient to occupy a campsite (thus, the nest) and send out hunters who could bring home meat, both killed or scavenged, to share with others. In exchange, the hunters received protection of the campsite and their own young offspring kept there.

From studies of modern humans, including hunter-gatherers, whose lives tell us so much about human origins, social psychologists have deduced the mental growth that began with hunting and campsites. A premium was placed on personal relationships geared to both competition and cooperation among the members. The process was ceaselessly dynamic and demanding. It far exceeded in intensity anything similar experienced by the wide-roaming, loosely organized bands of most animal societies. It required a memory good enough to assess the intentions of fellow members, as well as to predict their responses from one moment to the next, and, of decisive importance, it required the ability to invent and inwardly rehearse competing scenarios of future interactions.

The social intelligence of the campsite-anchored prehumans evolved as a kind of nonstop game of chess. Today, at the terminus of this evolutionary process, our immense memory banks are smoothly activated to join past, present, and future. They allow us to evaluate the prospects and consequences of alliances, bonding, sexual contact, rivalries, domination, deception, loyalty, and betrayal. We instinctively delight in the telling of countless stories about others, cast as players upon our own inner stage. The best of it is expressed in the creative arts, political theory, and other higher-level activities we have come to call the humanities.

The definitive part of the long creation story evidently began with the primitive *Homo habilis* (or a species closely related to it) two million years ago. Prior to the habilines the prehumans had been animals. Largely vegetarians, they had humanlike bodies, but their cranial capacity remained chimpanzee-sized, at or below 600 cubic centimeters (cc). Starting with the habiline period the capacity grew precipitously, to 680cc in *Homo habilis*,

900cc in *Homo erectus,* and about 1,400cc in *Homo sapiens.*
The expansion of the human brain was one of the most
rapid episodes of complex tissue evolution in the history
of life.

Yet to recognize the rare coming together of cooperat-
ing primates is not enough to account for the full potential of modern humans provided by a large brain capacity.
Evolutionary biologists have also searched for the grand
master of advanced social evolution, the combination of
forces and environmental circumstances that bestowed
greater longevity and more successful reproduction upon
the possessors of high social intelligence. Two competing theories of the principal force have been in conten-
tion. The first envisions kin selection: individuals favor
collateral kin (relatives other than offspring), making it
easier for altruism to evolve among members of the same
group. Complex social behavior can evolve when group
members individually reap greater benefits in numbers
of genes passed to the next generation than losses from
their altruism, averaged through their behavior toward
all members of the group. The combined effect on the
survival and reproduction of the individual is called
inclusive fitness, and the explanation of evolution by it is
called the theory of inclusive fitness.

In the second, more recently argued theory (full dis-
closure: I am one of the modern version's authors), the

grand master is multilevel selection. This formulation recognizes two levels at which natural selection operates: individual selection based on competition and cooperation among members of the same group, and group selection, which arises from competition and cooperation between groups. Group selection can occur through violent conflict or by competition between groups in the finding and harvesting of new resources. Multilevel selection is gaining in favor among evolutionary biologists because of recent mathematical proofs that kin selection can operate only under special conditions that rarely if ever exist. Also, multilevel selection is easily fitted to all of the known real animal cases of eusocial evolution, whereas kin selection, even when hypothetically plausible, can be fitted less well or not at all. (I'll treat this important subject in detail later, in Chapter 6.)

The roles of both individual and group selection are clear in the details of human social behavior. People are intensely interested in the minutiae of behavior of those around them. Gossip is a prevailing subject of conversation, everywhere from hunter-gatherer campsites to royal courts. The mind is a kaleidoscopically shifting map of others inside the group and a few outside, each of whom is evaluated emotionally in shades of trust, love, hatred, suspicion, admiration, envy, and sociability. We are compulsively driven to belong to groups or to create them as needed, which are variously nested, or overlapping, or separate, and in addition ranging from very large to very small. Almost all groups compete with those of similar kind in some manner or other. However gently expressed and generous in the tone of our discourse, we tend to think of our own group as superior, and we define our personal identities as members within them. The existence of competition, including military conflict, has been a hallmark of societies as far back in prehistory as archaeological evidence can be brought to bear.

The major features of the biological origins of *Homo sapiens* are coming into focus, and this clarification raises the potential of a more fruitful contact between science and the humanities. The convergence between these two great branches of learning will matter hugely when enough people have thought its potential through. On the science side, genetics as well as the brain sciences, evolutionary biology, and paleontology will each be seen in a different light. Students will be taught prehistory as well as conventional history, and the whole properly presented as the living world's greatest epic.

Pride and humility in better balance, we'll also take a more serious look at our place in nature. Exalted we are, risen to be the mind of the biosphere without a doubt, our spirits uniquely capable of awe and ever more breathtaking leaps of imagination. But we are still part of Earth's fauna and flora, bound to it by emotion, physiology, and, not least, deep history. It is folly to think of this planet as a way station to a better world. Equally, Earth would be unsustainable if converted into a literal, human-engineered spaceship.

Human existence may be simpler than we thought. There is no predestination, no unfathomed mystery of life. Demons and gods do not vie for our allegiance. Instead, we are self-made, independent, alone, and fragile, a biological species adapted to live in a biological world. What counts for long-term survival is intelligent self-understanding, based upon a greater independence of thought than that tolerated today even in our most advanced democratic societies.

3

Evolution and Our Inner Conflict

Are human beings intrinsically good but corruptible by the forces of evil, or the reverse, innately sinful yet redeemable by the forces of good? Are we built to pledge our lives to a group, even to the risk of death, or the opposite, built to place ourselves and our families above all else? Scientific evidence, a good part of it accumulated during the past twenty years, suggests that we are both of these things simultaneously. Each of us is inherently conflicted. Team player or whistle-blower? Charitable donation or personal certificates of deposit? Admitted traffic violation or denial? I don't believe I can let this subject pass by leaving my own conflicted emotions unconfessed. When Carl Sagan won the Pulitzer Prize for nonfiction in 1978, I dismissed it as a minor achievement for a scientist, scarcely worth listing. When I won the same prize the following year, it wondrously became a major literary award of which scientists should take special note.

We are all genetic chimeras, at once saints and sinners, champions of the truth and hypocrites—not because humanity has failed to reach some foreordained religious or ideological ideal, but because of the way our species originated across millions of years of biological evolution.

Don't get me wrong. I am not implying that we are driven by instinct in the manner of animals. Yet in order to understand the human condition, it is necessary to accept that we do have instincts, and it will be wise to take into account our very distant ancestors—as far back and in as fine a detail as possible. History alone cannot reach this level of understanding. It stops at the dawn of literacy, where it turns the rest of the story over to the detective work of archaeology. In still deeper time the quest becomes paleontology. For the real human story, history must comprise both the biological and cultural.

Within biology itself, the key to the mystery is the force that lifted prehuman social behavior to the human level. The leading candidate is multilevel selection, by which hereditary social behavior improves the competitive ability not just of individuals within groups but among groups as a whole.

Bear in mind that during organic evolution the unit

of natural selection is not the individual organism or the group, as some popular writers have misconstrued it. It is the gene (more precisely the alleles, or multiple forms of the same gene). The target of natural selection is the trait prescribed by the gene. The trait can be individual in nature and selected in competition among individuals inside or outside the group. Or the trait can be socially interactive in nature with other members of the group (as in communication and cooperation) and selected by competition among groups. A group of uncooperative, poorly communicating individuals will lose to its better organized competitors. The genes of the losers will decline across generations. Among animals, the consequences of group selection can be most plainly seen in the exquisitely programmed caste systems of ants, termites, and other social insects, but is also manifest in human societies. The idea of between-group selection as a force operating simultaneously in addition to between-individual selection is not new. Charles Darwin correctly deduced its role, first in the insects and then in human beings, respectively in *On the Origin of Species* and *The Descent of Man*.

I am convinced after years of research on the subject that multilevel selection, with a powerful role of group-to-group competition, has been a major force in the forging of advanced social behavior—including that of humans. In fact, it seems clear that so deeply ingrained are the evolutionary products of group-selected behaviors, so completely a part of the contemporary human condition are they, that we are prone to regard them as fixtures of nature, like air and water. They are instead idiosyncratic traits of our species. Among them is the intense, even obsessive interest of people in other people, which begins in the first days of life as infants learn particular scents and sounds of the adults around them. Research psychologists have found that all normal humans are geniuses at reading the intentions of others, whereby they evaluate, proselytize, bond, cooperate, gossip, and control. Each person, working his way back and forth through his social network, almost continuously reviews past experiences while imagining the consequences of future scenarios. Social intelligence of this kind occurs in many social animals, and reaches its highest level in chimpanzees and bonobos, our closest evolutionary cousins.

A second diagnostic hereditary trait of human behavior is the overpowering instinctual urge to belong to groups in the first place, shared with most kinds of social animals. To be kept forcibly in solitude is to be kept in pain, and put on the road to madness. A person's membership in his group—his tribe—is a large part of his identity. It also confers upon him to some degree or other a sense of superiority. When psychologists selected teams at random from a population of volunteers to compete in simple games, members of each team soon came to think of members of other teams as less able and trustworthy, even when the participants knew they had been selected at random.

All things being equal (fortunately things are seldom equal, not exactly), people prefer to be with others who look like them, speak the same dialect, and hold the same beliefs. An amplification of this evidently inborn predisposition leads with frightening ease to racism and religious bigotry. Then, also with frightening ease, good people do bad things. I know this truth from experience, having grown up in the Deep South during the 1930s and 1940s.

It might be supposed that the human condition is so distinctive and came so late in the history of life on Earth as to suggest the hand of a divine creator. Yet, as I've stressed, in a critical sense the human achievement was not unique at all. Biologists have identified at the time of this writing twenty evolutionary lines in the modern-world fauna that attained advanced social life based on some degree of altruistic division of labor. Most arose in the insects. Several were independent origins in marine shrimp, and three appeared among the mammals—that is, in two African mole rats, and us. All reached this level through the same narrow gateway: solitary individuals, or mated pairs, or small groups of individuals built nests and foraged from the nest for food with which they progressively raised their offspring to maturity.

Until about three million years ago the ancestors of *Homo sapiens* were mostly vegetarians, most likely wandering in groups from site to site where fruit, tubers, and other vegetable food could be harvested. Their brains were only slightly larger than those of modern chimpanzees. By no later than half a million years ago, however, groups of the ancestral species *Homo erectus* were maintaining campsites with controlled fire—the equivalent of nests—from which they foraged and returned with food, including a substantial portion of meat. Their brain size had increased to mid-sized, between that of chimpanzees and modern *Homo sapiens*. The trend appears to have begun one million to two million years previously, when the earlier prehuman ancestor *Homo habilis* turned increasingly to meat in its diet. With groups crowded together at a single site, and an advantage added by cooperative nest-building and hunting, social intelligence grew, along with the centers of memory and reasoning in the prefrontal cortex.

Probably at this point, during the habiline period, a

conflict ensued between individual-level selection, with individuals competing with other individuals in the same group, on the one side, and group-level selection, with competition among groups, on the other. The latter force promoted altruism and cooperation among all the group members. It led to innate group-wide morality and a sense of conscience and honor. The competition between the two forces can be succinctly expressed as follows: Within groups selfish individuals beat altruistic individuals, but groups of altruists beat groups of selfish individuals. Or, risking oversimplification, individual selection promoted sin, while group selection promoted virtue.

So it came to pass that humans are forever conflicted by their prehistory of multilevel selection. They are suspended in unstable and constantly changing positions between the two extreme forces that created us. We are unlikely to yield completely to either force as the ideal solution to our social and political turmoil. To give in completely to the instinctual urgings born from individual selection would be to dissolve society. At the opposite extreme, to surrender to the urgings from group selection would turn us into angelic robots—the outsized equivalents of ants.

The eternal conflict is not God's test of humanity. It is not a machination of Satan. It is just the way things worked out. The conflict might be the only way in the entire Universe that human-level intelligence and social organization can evolve. We will find a way eventually to live with our inborn turmoil, and perhaps find pleasure in viewing it as the primary source of our creativity.

II

THE UNITY OF KNOWLEDGE

 ALTHOUGH THE TWO GREAT BRANCHES OF

LEARNING, SCIENCE AND THE HUMANITIES, ARE

RADICALLY DIFFERENT IN THE WAY THEY DESCRIBE

OUR SPECIES, THEY HAVE RISEN FROM THE SAME

WELLSPRING OF CREATIVE THOUGHT.

4

The New Enlightenment

We've considered thus far the biological origins of human nature, and from this information the idea that a large part of human creativity is generated by the inevitable and necessary conflict between the individual and group levels of natural selection. The implied unity in the explanation leads us to the next leg of the journey I suggest. It is the concept that science and the humanities share the same foundation, in particular that the laws of physical cause and effect can somehow ultimately account for both. You will likely recognize this proposition. Western culture has already traveled this way. It was called the Enlightenment.

During the seventeenth and eighteenth centuries, the idea of the Enlightenment ruled the Western intellectual world. At that time it was a juggernaut; in the minds of many it even seemed to be the destiny of the human species. Scholars appeared on track to explain both the Universe and the meaning of humanity by the laws of science, the latter called at that time natural philosophy. The great branches of learning, Enlightenment scholars believed, can be unified by a continuous network of cause and effect. Then, when built from reality and reason alone, cleansed of superstition, all of knowledge might come together to form what in 1620 Francis Bacon, greatest of the Enlightenment's forerunners, termed "the empire of man."

The Enlightenment quest was driven by the belief that entirely on their own, human beings can know all that needs to be known, and in knowing understand, and in understanding gain the power to choose more wisely than ever before.

By the early 1800s, however, the dream faltered and Bacon's empire retreated. There were two reasons. First, although scientists were generating discoveries at an exponential pace, they were nowhere close to meeting the expectations of the more optimistic Enlightenment thinkers. Second, this shortfall allowed the founders of the Romantic tradition of literature, including some of the greatest poets of all time, to reject the presumptions of the Enlightenment worldview and seek meaning in other, more private venues. Science had no way, and it never would, to touch what people deeply feel and express only through the creative arts. Reliance on scientific knowledge, many believed and their contemporary successors continue to believe, beggars the human potential.

For the next two centuries and to the present day, science and the humanities went their own ways. Physicists of course no less continue to enjoy playing in string quartets, and novelists write books that marvel at the wonders uncovered by science. But the two cultures—as they came to be called by the middle of the twentieth century—were considered by most to be separated by a permanent chasm built into the mind, perhaps intrinsic to the nature of existence itself.

In any case there was simply no time during the long eclipse of the Enlightenment to think of unification. In order to accommodate the rising flood of information, scientific disciplines were dividing into specialties at a near-bacterial rate—fast then faster and then even faster. The creative arts for their part continued to flower with brilliant and idiosyncratic expressions of the human imagination. There was very little interest in trying to reignite what was perceived as an antique and hopeless philosophical quest. Yet the Enlightenment was never proved to be impossible. It was not dead. It was just stalled.

Is there any value in resuming the quest now, and any chance of achieving it? Yes, because enough is known today to make it more attainable than during its first flowering. And yes, because the solutions of so many problems of modern life hinge on solutions for the clash of competing religions, the ambiguities of moral reasoning, the inadequate foundations of environmentalism, and (the big one) the meaning of humanity itself.

Studying the relation between science and the humanities should be at the heart of liberal education everywhere, for students of science and the humanities alike. That's not going to be easy to achieve, of course. Among the fiefdoms of academia and punditry there exists a great variation in acceptable ideology and procedure. Western intellectual life is ruled by hard-core specialists. At Harvard University, for example, where I taught for four decades, the dominant criterion in the selection of new faculty was preeminence or the promise of preeminence in a specialty. Starting with the deliberations of department-level search committees, then recommendations to the dean of the faculty of arts and sciences, and at last the final decision by the president of Harvard, who was assisted by an ad hoc committee drawn from both within and outside the university, the pivotal question asked was, "Is the candidate the best in the world in his research specialty?" On teaching, it was almost always an easygoing, "Is the candidate adequate?" The guiding philosophy overall was that the assembly of a sufficient number of such world-class specialists would somehow coalesce into an intellectual superorganism attractive to both students and financial backers.

The early stages of a creative thought, the ones that count, do not arise from jigsaw puzzles of specialization. The most successful scientist thinks like a poet— wide-ranging, sometimes fantastical—and works like a bookkeeper. It is the latter role that the world sees. When writing a report for a technical journal or speaking at a conference of fellow specialists, the scientist avoids metaphor. He is careful never to be accused of rhetoric or poetry. A very few loaded words may be used, if kept to the introductory paragraphs and the discussion following the presentation of data, and if added to clarify the meaning of a technical concept, but they are never used for the primary purpose of stirring emotion. The language of the author must at all times be restrained and obedient to logic based on demonstrable fact.

The exact opposite is the case in poetry and the other creative arts. There metaphor is everything. The creative writer, composer, or visual artist conveys, often obliquely by abstraction or deliberate distortion, his own perceptions and the feelings he hopes to evoke—about something, about anything, real or imagined. He seeks to bring forth in an original way some truth or other about the human experience. He tries to pass what he creates directly along the channel of human experience, from his mind to your mind. His work is judged by the power and beauty of its metaphors. He obeys a dictum ascribed to Picasso: art is the lie that shows us the truth.

Wildly searching, sometimes shocking in effect, the creative arts and much of the humanities scholarship analyzing them are nonetheless in an important sense just the same old story, with the same themes, the same archetypes, the same emotions. We readers don't care. We're addicted to anthropocentricity, bound to a bottomless fascination with ourselves and others of our kind. Even the best-educated live on an ad libitum diet of novels, movies, concerts, sports events, and gossip all designed to stir one or more of the relatively small range of emotions that diagnose *Homo sapiens*. Our stories about animals require human-like emotions and behavior understandable with well-worn guidebooks of human nature. We use endearing animal caricatures, including even those of tigers and other ferocious predators, to teach children about other people.

We are an insatiably curious species—provided the subjects are our personal selves and people we know or would like to know. The behavior goes far back beyond our species in the evolution of the primate family tree. It has been observed, for example, that when caged monkeys are allowed to look outside at a variety of other objects, their first choice for attention is other monkeys.

The function of anthropocentricity—fascination about ourselves—is the sharpening of social intelligence, a skill in which human beings are the geniuses among all Earth's species. It arose dramatically in concert with the evolution of the cerebral cortex during the origin of *Homo sapiens* from the African australopith prehumans. Gossip, celebrity worship, biographies, novels, war stories, and sports are the stuff of modern culture because a state of intense, even obsessive concentration on others has always enhanced survival of individuals and groups. We are devoted to stories because that is how the mind works—a never-ending wandering through past scenarios and through alternative scenarios of the future.

If gods of the ancient Greek tradition were watching, they would view human error the way we do in comedies and tragedies, but they might also feel empathy by viewing our foibles as flaws forced on us by Darwinian necessity. There is a parallel to gods and their human puppets in people watching kittens at play. The animals use three basic maneuvers suited to their future role as predators: to stalk and leap on a trailing string is practice for catching mice; to leap up to the string above and seize it with paws clapped together is for birds; and to scoop at a string near the feet is for fish or small prey at their feet.

It's all amusing to us, but vital to them as a sharpening of survival skills.

Science builds and tests competitive hypotheses from partial evidence and imagination in order to generate knowledge about the real world. It is totally committed to fact without reference to religion or ideology. It cuts paths through the fever swamp of human existence.

You have of course heard of these qualities. But science has additional properties that distinguish it from the humanities. Of these the most important is the concept of the continuum. The idea of variation of entity and process occurring continuously in one, two, or more dimensions is so routine in most of physics and chemistry as to require no explicit mention. Continua include such familiar gradients as temperature, velocity, mass, wave length, particle spin, pH, and carbon-based molecular analogs. They become less obvious in molecular biology, where only a few basic variations in structure work to explain the function and reproduction of cells. They reappear forcefully in evolutionary biology and evolution-based ecology, which address the differing adaptations of millions of species to their respective environments. And they have returned with even greater flair and drama in studies of exoplanets.

Some nine hundred such planets had been discovered prior to the partial shutdown of the Kepler space telescope in 2013, the shutdown due to the malfunction of an aiming device. The Kepler images were amazing even to generations who regarded flybys and soft landings on other planets in the Solar System as routine. They are also immensely important, the equivalent of a seaman's first glimpse of a new continent's coastline, and a shout of, Land! Land!, where none might have existed. An estimated hundred billion star systems make up the Milky Way galaxy, and astronomers believe that all are orbited by an average of at least one planet. A small but still substantial fraction are likely to harbor life-forms—even if the organisms are only microbes living under extremely hostile conditions.

The exoplanets (planets in other star systems) of the galaxy form a continuum. Astronomers have newly observed or at least inferred a bestiary of exoplanets more varied than anything previously imagined. There exist giant gas planets resembling Jupiter and Saturn, some hugely larger in volume. There are smaller rocky planets like our own, tiny specks orbiting at the right distance from the mother star to support life, fundamentally different from rocky planets at other distances (as Mercury and Venus are fatally near to the Sun and planetlike Pluto fatally far away). There exist planets that do not rotate, others that travel close to the mother star and then far away and back again in elliptical orbits.

There probably exist orphaned rogue planets, thrown loose from the gravitational pull of their mother stars, drifting through outer space. Some of the exoplanets also have an entourage of one or more moons. In addition to great and continuous variation in size, location, and orbit, there are comparable gradients in the chemical composition of the body and atmosphere of the planets and their moons, derived from the particularities of their origin.

Astronomers, being normal humans as well as scientists, are as awed as the rest of us by their discoveries. The discoveries affirm that Earth is not the center of the Universe—we've known that since Copernicus and Galileo—but just how far from the center has been hard to imagine. The tiny blue speck we call home is proportionately no more than that, a mote of stardust near the edge of our galaxy among a hundred billion or more galaxies in the universe. It occupies just one position in a continuum of planets, moons, and other planetlike heavenly bodies that we have just begun to understand. It would be becoming of us to speak modestly of our status in the cosmos. Let me offer a metaphor: Earth relates to the Universe as the second segment of the left antenna of an aphid sitting on a flower petal in a garden in Teaneck, New Jersey, for a few hours this afternoon.

With botany and entomology thus fleetingly brought

to mind, it is appropriate to introduce another continuum, the diversity of life in Earth's biosphere. At the
time of this writing (in 2013) there are 273,000 known
species of plants in the living flora of Earth, a number
expected to rise to 300,000 as more expeditions take
to the field. The number of all known species of organisms on Earth, plants, animals, fungi, and microbes, is
about 2 million. The actual number, combining known
and unknown, is estimated to be at least three times that
number, or more. The roster of newly described species
is about 20,000 a year. The rate will certainly grow, as
a multitude of still poorly explored tropical forest fragments, coral reefs, seamounts, and uncharted ridges and
canyons of the deep ocean floor become better known.
The number of described species will accelerate even
faster with exploration of the largely unknown microbial world, now that the technology needed for the
study of extremely small organisms has become routine.
There will come to light strange new bacteria, archaeans,
viruses, and picozoans that still swarm unseen everywhere on the surface of the planet.

As the census of species proceeds, other continua of
biodiversity are being mapped. They include the unique
biology of each living species and the long, winding processes of evolution that created it. Part of the end product
is the gradient of size across a dozen orders of magni-

tude. It ranges from the blue whale and African elephant down to superabundant photosynthetic bacteria and scavenging picozoans of the sea, the latter so small they cannot be studied with ordinary light microscopy.

Of all the continua mapped by science, the most relevant to the humanities are the senses, which are extremely limited in our species. Vision is based in *Homo sapiens* on an almost infinitesimal sliver of energy, four hundred to seven hundred nanometers in the electromagnetic spectrum. The rest of the spectrum, saturating the Universe, ranges from gamma rays trillions of times shorter than the human visual segment to radio waves trillions of times longer. Animals live within their own slivers of continua. Below four hundred nanometers, for example, butterflies find pollen and nectar in flowers by the patterns of ultraviolet light reflected off the petals—patterns and colors unseen by us. Where we see a yellow or red blossom, the insects see an array of spots and concentric circles in light and dark.

Healthy people believe intuitively that they can hear almost every sound. However, our species is programmed to detect only twenty to twenty thousand hertz (cycles of air compression per second). Above that range, flying bats broadcast ultrasonic pulses into the night air and listen for the echoes to dodge obstacles and snatch moths and other insects on the wing. Below the human range, elephants rumble complex messages in exchanges back and forth with other members of their herd. We walk through nature like a deaf person on the streets of New York, sensing only a few vibrations, able to interpret almost nothing.

Human beings have one of the poorest senses of smell of all the organisms on Earth, so weak that we have only a tiny vocabulary to express it. We depend heavily on similes such as "lemony" or "acidic" or "fetid." In contrast, the vast majority of other organisms, ranging in kind from bacteria to snakes and wolves, rely on odor and taste for their very existence. We depend on the sophistication of trained dogs to lead us through the olfactory world, tracking individual people, detecting even the slightest trace of explosives and other dangerous chemicals.

Our species is almost wholly unconscious of certain other kinds of stimuli without the use of instruments. We detect electricity solely by a tingle, a shock, or a flash of light. In contrast, there exist a variety of freshwater eels, catfish, and elephant-nose fish, confined to murky water where, deprived of vision, they live instead in a galvanic world. They generate charged fields around their bodies with trunk muscle tissue that has been modified by evolution into organic batteries. With the aid of electric shadows in the pattern of charges, the fish avoid obstacles around them, locate prey, and communicate with others of the same species. Yet another part of the environment beyond the reach of humans is Earth's magnetic field, used by some migratory birds to guide them during their long-distance journeys.

The exploration of continua allows humanity to measure the dimensions of the real cosmos, from the infinite ranges of size, distance, and quantity, in which we and our little planet exist. The scientific enterprise suggests where to look for previously unexpected phenomena, and how to perceive the whole of reality by a measurable webwork of cause-and-effect explanation. By knowing the position of each phenomenon in the relevant continua—relevant continua in ordinary parlance being the variable of each system—we have learned the chemistry of the surface of Mars; we know approximately how and when the first tetrapods crawled out of ponds onto the land; we can predict conditions in both the infinitesimal and near-infinite by the unified theory of physics; and we can watch blood flow and nerve cells in the human brain light up during conscious thought. In time, likely no more than several decades, we will be able to explain the dark matter of the Universe, the origin of life on Earth, and the physical basis of human consciousness during changes of mood and thought. The invisible is seen, the vanishingly small weighed.

So, what has this explosive growth of scientific knowledge to do with the humanities? *Everything.* Science and technology reveal with increasing precision the place of humanity, here on Earth and beyond in the cosmos as a whole. We occupy a microscopic space in each of the relevant continua that might have produced a species of human-grade intelligence anywhere, here and on other planets. Our ancestral species, traced further and further back through a series of ever more primitive life-forms, are all lucky lottery winners that stumbled their way through the labyrinth of evolution.

We are a very special species, perhaps the chosen species if you prefer, but the humanities by themselves cannot explain why this is the case. They don't even pose the question in a manner that can be answered. Confined to a small box of awareness, they celebrate the tiny segments of the continua they know, in minute detail and over and over again in endless permutations. These segments alone do not address the origins of the traits we fundamentally possess—our overbearing instincts, our moderate intelligence, our dangerously limited wisdom, even, critics will insist, the hubris of our science.

The first Enlightenment was undertaken more than four centuries ago when science and the humanities both were elementary enough to make their symbiosis look feasible. It became possible with the opening of the global sea routes by Western Europe from the late fifteenth century onward. The circumnavigation of Africa and the discovery of the New World led to new, global trade routes and expanded military conquest. The new, global reach was a turning point in history that placed a premium on knowledge and invention. Now we are launched into a new cycle of exploration—infinitely richer, correspondingly more challenging, and not by coincidence increasingly humanitarian. It is within the power of the humanities and the serious creative arts within them to express our existence in ways that begin at last to realize the dreams of the Enlightenment.

5

The All-Importance
of the Humanities

You might think this odd coming from a data-driven biologist, but I believe that the extraterrestrials created by the confabulations of science fiction serve us in an important way: they improve reflection on our own condition. When made as fully plausible as science allows, they help us to predict the future. Real aliens would tell us, I believe, that our species possesses one vital possession worthy of their attention. It is not our science and technology, as you might think. It is the humanities.

These imagined yet plausible aliens have no desire to please or elevate our species. Their relation to us is benevolent, the same as our own toward wildlife grazing and stalking in the Serengeti. Their mission is to learn all they can from the singular species that achieved civilization on this planet. Wouldn't that have to be the secrets of our science? No, not at all. We have nothing to teach them. Keep in mind that nearly everything that can be called science is less than five centuries old. Because scientific knowledge has been more or less doubling according to discipline (such as physical chemistry and cell biology) every one or two decades for the past two centuries, it follows that what we know is by geological standards brand-new. Technological applications are also in an early stage of evolution. Humanity entered our present global, hyperconnected technoscientific era only two decades ago—less than an eyeblink in the starry message of the cosmos. By chance alone, and given the multibillion-year age of the galaxy, the aliens reached our present-day, still-infantile level millions of years ago. It could have been as much as a hundred million years ago. What then can we teach our extraterrestrial visitors? Put another way, what could Einstein as a toddler have taught a professor of physics? Nothing at all. For the same reason our technology would be vastly inferior. If that were not so, we would be the extraterrestrial visitors and they the planetary aboriginals.

So what could the hypothetical aliens learn from us that has any value to them? The correct answer is the humanities. As Murray Gell-Mann once remarked of the field he has pioneered, theoretical physics consists of a small number of laws and a great many accidents. The same is true, a fortiori, of all the sciences. The origin of life occurred over three and a half billion years ago. The subsequent diversification of the primordial organisms into species of microbes, fungi, plants, and animals is only one history that could have occurred out of a near-infinitude of histories. The extraterrestrial visitors would know this, from robot probes and the principles of evolutionary biology. They could not immediately fathom Earth's full history of organic evolution, with its extinctions, replacements, and dynastic rise and fall of major groups—cycads, ammonoids, dinosaurs. But with their super-efficient fieldwork and DNA-sequencing and proteonomic technology, they would quickly learn Earth's fauna and flora at the present moment, and the nature and ages of the forerunners, and calculate patterns in space and time of life's evolutionary history. It's all a matter of science. The aliens would soon know all that we know called science, and much more, as though we had never existed.

In a closely parallel manner during the human history of the past hundred thousand years or so, a small number of human Ur-cultures arose, then gave birth to the thousands of daughter cultures. Many of these persist today, each with its one language or dialect, religious beliefs, and social and economic practices. Like species of plants and animals splintering across the geological ages, they have continued to evolve, alone, or divided into two more cultures, perhaps fused in part, and some have just disappeared. Of the nearly seven thousand languages currently spoken worldwide, 28 percent are used by fewer than a thousand people, and 473 are on the edge of extinction, spoken only by a handful of elderly people. Measured this way, recorded history and prehistory before it present a kaleidoscopic pattern similar to that of species formation during organic evolution—yet different in major ways from it.

Cultural evolution is different because it is entirely a product of the human brain, an organ that evolved during prehuman and Paleolithic times through a very special form of natural selection called gene-culture coevolution (where genetic evolution and cultural evolution each affect the trajectory of the other). The brain's unique capability, lodged primarily in the memory banks of the frontal cortex, arose from the tenure of *Homo habilis* two million to three million years ago until the global spread of its descendant *Homo sapiens* sixty thousand years ago. To understand cultural evolution from the outside looking in, as opposed to the inside looking out, the way we do it, requires interpreting all of the intricate feelings and constructions of the human mind. It requires intimate contact with people and knowledge of countless personal histories. It describes the way a thought is translated into a symbol or artifact. All this the humanities do. They are the natural history of culture, and our most private and precious heritage.

There is another cardinal reason for treasuring the humanities. Scientific discovery and technological advance have a life cycle. In time, after reaching an immense size and unimaginable complexity, they will certainly slow and stabilize at a much lower level of growth. Within the span of my own career as a published scientist across half a century, the number of discoveries per researcher per year has declined dramatically. Teams have grown larger, with ten or more coauthors on technical papers now a commonplace. The technology required to make a scientific discovery in most disciplines has become much more complex and expensive, and the new technology and statistical analysis required for scientific research more advanced.

Not to worry. By the time the process has set in, likely in this century, the role of science and high technology will, as expected, be beneficent and far more pervasive than now. But—and this is the most important part— science and technology will also be the same everywhere, for every civilized culture, subculture, and person. Sweden, the United States, Bhutan, and Zimbabwe will share the same information. What will continue to evolve and diversify almost infinitely are the humanities.

For the next few decades, most major technological advances are likely to occur in what is often denoted BNR: biotechnology, nanotechnology, and robotics. In pure science the secular grails now sought along the broad frontier include the deduction of how life originated on Earth, along with the creation of artificial organisms, gene substitution and surgically precise modification of the genome, discovery of the physical nature of consciousness, and, not least, the construction of robots that can think faster and work more efficiently than humans in most blue-collar and white-collar labor. At the present time these envisioned advances are the stuff of science fiction. But not for long. Within a few decades they will be reality.

And the cards are now on the table, faceup. First on the agenda is the correction of the more than a thousand genes for which rare mutant alleles have been identified as the cause of hereditary diseases. The method of choice will be gene substitution, replacing the mutant allele with a normal one. Although still in the earliest, mostly untested stage, it promises eventually to replace amniocentesis, which allows first a readout of the embryonic chromosome structure and genetic code, then therapeutic abortion to avoid disability or death. Many people object to therapeutic abortions, but I doubt that many would object to gene substitution, which can be compared with replacement of a defective heart valve or diseased kidney.

An even more advanced form of a volitional evolution, albeit indirect in cause, is the homogenization ongoing among the world populations by increased emigration and interracial marriage. The result is a massive redistribution of *Homo sapiens* genes. Genetic variation between populations is declining, genetic variation within populations is increasing, and, as an overall result, the genetic variation of the species as a whole is also increasing—the last dramatically so. These trends create a dilemma of volitional evolution likely to catch the attention of even the most myopic political think tanks in a few decades. Do we wish to guide the evolution of diversity in order to increase the frequency of desirable traits? Or increase it still more? Or finally—this will almost certainly be the short-term decision—just leave it alone and hope for the best?

Such alternatives are not science fiction, and they are not frivolous. On the contrary, they are linked to yet another biology-based dilemma that has already entered public discussion, ranking with contraceptives in high school and evolution-free textbooks in Texas. It is this: With more and more decision making and work done by robots, what will be left for humans to do? Do we really want to compete biologically with robot technology by using brain implants and genetically improved intelligence and social behavior? This choice would mean a sharp departure away from the human nature we have inherited, and a fundamental change in the human condition.

Now we are talking about a problem best solved within the humanities, and one more reason the humanities are all-important. While I'm at it, I hereby cast a vote for existential conservatism, the preservation of biological human nature as a sacred trust. We are doing very well in science and technology. Let's agree to keep it up, and move both along even faster. But let's also promote the humanities, that which makes us human, and not use science to mess around with the wellspring of this, the absolute and unique potential of the human future.

6

The Driving Force
of Social Evolution

Few questions in biology are as important as the evolutionary origin of instinctive social behavior. To find the correct answer is to explain one of the great transitions in levels of biological organization, from the organism to the superorganism—from one ant, say, to an organized colony of ants, and from a solitary primate to an organized society of human beings.

The most complex forms of social organization are made from high levels of cooperation. They are furthered with altruistic acts performed by at least some of the colony members. The highest level of cooperation and altruism is that of eusociality, in which some colony members surrender part or all of their personal reproduction in order to increase reproduction by the "royal" caste specialized for that purpose.

As I've pointed out, there are two competing theories of the origin of advanced social organization.

One is the standard theory of natural selection. It has proved correct across a broad range of social and nonsocial phenomena, improving in precision since the origin of modern population genetics in the 1920s and modern synthesis of evolutionary theory in the 1930s. It is based on the principle that the unit of heredity is the gene, which typically acts as part of a network of genes, and the target of natural selection is the trait prescribed by the gene. For example, an unfavorable mutant gene in humans is that which prescribes cystic fibrosis. The gene is rare because its phenotype cystic fibrosis is selected against—it lowers longevity and reproduction. Examples of favorable mutant genes are those that prescribe adult lactose tolerance. After originating in dairying populations in Europe and Africa, the phenotype prescribed by the mutated genes made milk available as a reliable adult food, and thereby increased the comparative longevity and reproduction of the people possessing them.

A gene for a trait that affects a group member's longevity and reproduction relative to other members in the same group is said to be subject to individual-level natural selection. A gene for a trait entailing cooperation and other forces of interaction with fellow group members may or may not be subject to individual-level selection. In either case it is also likely to affect longevity and reproduction of the group as a whole. Because groups compete with other groups, in both conflict and their relative efficiency in resource extraction, their differing traits are subject to natural selection. In particular, the genes prescribing interactive (hence social) traits are subject to group-level selection.

Here is a simplified scenario of evolution according to the standard theory of natural selection. A successful thief furthers his own interests and those of his offspring, but his actions weaken the remainder of the group. Any genes proscribing his psychopathic behavior will increase within the group from one generation to the next—but, like a parasite causing a disease in an organism, his activity weakens the rest of the group—and eventually the thief himself. At the opposite extreme, a valiant warrior leads his group to victory, but in doing so is killed in battle, leaving few or no offspring. His genes for heroism are lost with him, but the remainder of the group, and the heroism genes they share, benefit and increase.

The two levels of natural selection, individual and group, illustrated by these extremes, are in opposition. They will in time lead to either a balance of the opposing genes or an extinction of one of the two kinds altogether. Their action is summarized in this maxim: selfish members win within groups, but groups of altruists best groups of selfish members.

The theory of inclusive fitness, in opposition to the standard theory of natural selection, and with it the established principles of population genetics, treats the individual group member, not its individual genes, as the unit of selection. Social evolution arises from the sum of all the interactions of the individual with each of the other group members in turn, multiplied by the degree of hereditary kinship between each pair. All the effects of this multiplicity of interactions on the individual, both positive and negative, make up its inclusive fitness.

Although the controversy between natural selection and inclusive fitness still flickers here and there, the assumptions of the theory of inclusive fitness have proved to be applicable only in a few extreme cases unlikely to occur on Earth or any other planet. No example of inclusive fitness has been directly measured. All that has been accomplished is an indirect analysis called the regressive method, which unfortunately has itself been mathematically invalidated. The use of the individual or group as the unit of heredity, rather than the gene, is an even more fundamental error.

At this point, prior to developing the theories further, it will be instructive to take a specific example in the evolution of social behavior and see how it is treated respectively by each approach.

The life cycle of ants has always been a favorite of inclusive fitness theorists as offering proof of the role of kinship and the validity of inclusive fitness. Many ant species have the following life cycle: their colonies reproduce by releasing virgin queens and males from the nest. After mating, the queens do not return home, but disperse to establish new colonies on their own. The males die within hours. The virgin queens are much larger than the males, and colonies invest a correspondingly larger fraction of their resources to their production.

The inclusive fitness explanation of the size difference between the sexes, introduced in the 1970s by the biologist Robert Trivers, is as follows. The means of sex determination in ants is peculiar, such that sisters are more closely related to one another than they are to their brothers (providing the queens mate with only one male). Because the workers raise the young, Trivers continued, and because they favor sisters over brothers, they invest more in virgin queens than in males. The colony, with workers in control, accomplish this end by making the queens individually much larger in size. This process deduced with inclusive fitness theory is called indirect natural selection.

The standard population genetics model, in contrast, posits direct natural selection and tests it with direct observation in the field and laboratory. The larger size of the virgin queen is necessary, as all entomologists know, because of the way she starts a new colony. She digs a nest, seals herself in, and raises the first brood of workers on her large bodily reserves of fat and metabolized wing muscles. The male is small because its only function is to mate. After achieving insemination, it dies. (Queens live on in a few species, incidentally, for more than twenty years.) The roundabout inclusive fitness explanation for investments according to gender is therefore wrong.

The assumption of inclusive fitness theory that workers control the colony's allocation, a crucial point in this reasoning, is also wrong. Using the valve on her spermatheca, the bag-like organ in which the sperm are stored, the queen determines the sex of the offspring born. If a sperm is released to fertilize an egg in the queen's ovary, a female is born. If no sperm is released, the egg is not fertilized, and from the unfertilized egg a male is born. Thereafter, a complex of factors, only some of which are under worker control, determine which female eggs and larvae will become queens.

For half a century, while data were still relatively scarce, the theory of inclusive fitness was the prevailing explanation of the origin of advanced social behavior. It began in 1955 with a simple mathematical model by the British geneticist J. B. S. Haldane. His argument was in the following form (which I've altered here a bit to make it intuitively easier). Imagine that you are a child-less bachelor standing on a riverbank. Looking out over the water, you see that your brother has fallen in and is drowning. The river that day is raging, and you're a poor swimmer, so you know that if you jump in and save him, you yourself will probably drown. So the rescue requires altruism on your part. But (Haldane said) it does not also require altruism on the part of your genes, including those responsible for making you altruistic. The reason is the following. Because the man is your brother, half of his genes are identical to yours. So you jump in, save him, and sure enough, you drown. Now you're gone, but half of your genes are saved. All your brother has to do in order to make up the loss in genes is to have two addi-tional children. The genes are the unit of selection; the genes are what count in evolution by natural selection.

In 1964, another British geneticist, William D. Ham-ilton, expressed Haldane's concept in a general formula, which came to be known in later years as the Hamilton inequality. It said that a gene prescribing altruism, such as that of the heroic brother, will increase if the bene-fit in number of offspring to the recipient exceeds the cost in offspring to the altruist. However, this advan-tage to the altruist will be effective only if the recipient and the altruist are closely related. The degree of kin-ship is the fraction of genes that are shared by the altruist and recipient due to their common descent: one-half between siblings, one-eighth between first cousins, and so on in a rapidly declining rate as the degree of kinship becomes more distant. The process later came to be called kin selection. It seemed, at least from this line of reasoning, that close kinship is the key to the biological origin of altruism and cooperation. Hence close kinship is a primary factor of advanced social evolution.

On the surface, kin selection seemed at first to be a reasonable explanation for the origin of organized societies. Consider any group of individuals that have come together in one manner or another but remain unorganized—a fish school, for example, a flock of birds, or a local population of ground squirrels. The group members, let us say, are able to distinguish not just their own offspring, leading to evolution of parental care by standard (Darwinian) natural selection. Suppose they also recognize collateral relatives related by common descent such as siblings and cousins. Allow further that mutations occur that induce individuals to favor close collateral relatives over distant relatives or non-relatives. An extreme case would be Haldane's heroism biased toward a brother. The result would be nepotism, resulting in a Darwinian advantage over others in the group. But where does that lead an evolving population? As the collateral-favoring genes spread, the group would change into an ensemble not of competing individuals and their offspring, but of an ensemble of parallel competing extended families. To achieve group-wide altruism, cooperation, and division of labor, in other words organized societies, requires a different level of natural selection. That level is group selection.

Also in 1964, Hamilton took the kinship principle one step further by introducing the concept of inclusive fitness. The social individual lives in a group, and it interacts with other members of the group. The individual participates in kin selection with each of the other group members with which it interacts. The added effect it has on its own genes passed into the next generation is its inclusive fitness: the sum of all the benefits and costs, discounted by the degree of kinship with each other group member. With inclusive fitness the unit of selection had passed subtly from the gene to the individual.

At first I found the theory of inclusive fitness, winnowed down to a few cases of kin selection that might be studied in nature, enchanting. In 1965, a year after Hamilton's article, I defended the theory at a meeting of the Royal Entomological Society of London. Hamilton himself was at my side that evening. In my two books formulating the new discipline of sociobiology, *The Insect Societies* (1971) and *Sociobiology: The New Synthesis* (1975), I promoted kin selection as a key part of the genetic explanation of advanced social behavior, treating it as equal in importance to caste, communication, and the other principal subjects that make up sociobiology. In 1976 the eloquent science journalist Richard Dawkins explained the idea to the general public in his best-selling book *The Selfish Gene*. Soon kin selection and some version of inclusive fitness were installed in textbooks and popular articles on social evolution. During the following three decades a large volume of general and abstract extensions of the theory of kin selection was tested, especially in ants and other social insects, and purportedly found proof in studies on rank orders, conflict, and gender investment.

By 2000 the central role of kin selection and its extensive inclusive fitness had approached the stature of dogma. It was a common practice for writers of technical papers to acknowledge the truth of the theory, even if the content of the data to be presented were only distantly relevant to it. Academic careers had been built upon it by then, and international prizes awarded.

Yet the theory of inclusive fitness was not just wrong, but fundamentally wrong. Looking back today, it is apparent that by the 1990s two seismic flaws had already appeared and begun to widen. Extensions of the theory itself were growing increasingly abstract, hence remote from the empirical work that continued to flourish elsewhere in sociobiology. At the same time the empirical research devoted to the theory remained limited to a small number of measurable phenomena. Writings on the theory mostly in the social insects were repetitive. They offered more and more about proportionately fewer topics. The grand patterns of ecology, phylogeny, division of labor, neurobiology, communication, and social physiology remained virtually untouched by the asseverations of the inclusive theorists. Much of the popular writing devoted to it was not new but affirmative in tone, declaring how great the theory was yet to become.

Inclusive fitness theory, fondly called IF theory for short by its defenders, was showing increasing signs of senescence. By 2005 questions about its soundness were being openly expressed, especially among leading experts on the details of the biology of the ants, termites, and other eusocial insects, as well as a few theoreticians bold enough to seek alternative explanations of the origin and evolution of eusociality. The researchers most committed to IF theory either ignored these deviations or summarily dismissed them. By 2005 they had gained enough representation in the anonymous peer review system to hinder publication of contrary evidence and opinions in leading journals. For example, a keystone early support of inclusive fitness theory, cited in textbooks, was the prediction of overrepresentation of the

Hymenoptera (bees, wasps, ants) among eusocial animal species. When after a time one investigator pointed out that new discoveries had nullified the prediction, he was told, in effect, "We already knew that." They did know that, but hadn't done more than just drop the subject. The "Hymenoptera hypothesis" was not wrong; it had just become "irrelevant." When a senior investigator used field and laboratory studies to show that primitive termite colonies compete with one another and grow in part by the fusion of unrelated workers, the data were rejected on grounds that the conclusion did not adequately take into account inclusive fitness theory.

Why did an outwardly arcane topic of theoretical biology excite such fierce partisanship? Because the problem it addresses is of fundamental importance, and the stakes in trying to solve it had become exceptionally high. Furthermore, inclusive fitness was beginning to resemble a house of cards. To pull even one out risked collapsing the whole. Pulling cards, however, was worth the price to reputation. There existed in the air the promise of a paradigm shift, a rare event in evolutionary biology.

In 2010, the dominance of inclusive fitness theory was finally broken. After struggling as a member of the small but still muted contrarian school for a decade, I joined two Harvard mathematicians and theoretical biologists, Martin Nowak and Corina Tarnita, for a top-to-bottom analysis of inclusive fitness. Nowak and Tarnita had independently discovered that the foundational assumptions of inclusive fitness theory were unsound, while I had demonstrated that the field data used to support the theory could be explained equally well, or better, with direct natural selection—as in the sex-allocation case of ants just described.

Our joint report was published on August 26, 2010, as the cover article of the prestigious journal *Nature*. Knowing the controversy involved, the *Nature* editors had proceeded with unusual caution. One of them familiar with the subject and the mode of mathematical analysis came from London to Harvard to hold a special meeting with Nowak, Tarnita, and myself. He approved, and the manuscript was next examined by three anonymous experts. Its appearance, as we expected, caused a Vesuvian explosion of protest—the kind cherished by journalists. No fewer than 137 biologists committed to inclusive fitness theory in their research or teaching signed a protest in a *Nature* article published the following year. When I repeated part of my argument as a chapter in the 2012 book *The Social Conquest of Earth*, Richard Dawkins responded with the indignant fervor of a true believer. In his review for the British magazine *Prospect*, he urged others not to read what I had written, but instead to cast the entire book away, "with great force," no less.

Yet no one since that time has refuted the mathematical analysis by Nowak and Tarnita, or my argument favoring the standard theory over inclusive fitness theory in the interpretation of field data.

In 2013 Nowak and I were joined by another mathematical biologist, Benjamin Allen, in a still deeper expansion of the ongoing analysis. (Tarnita had moved to Princeton, where she was busy adding field research to her mathematical modeling.) In late 2013 we published the first in a planned series of refereed articles. Because of the need for exactitude, and the material that these articles contain that may be relevant to the history and philosophy of the subject, I've taken the step of providing a simplified summary of the first one in the appendix of this book.

Now at last we can return to a key question in a more open spirit of inquiry: What was the driving force in the origin of human social behavior? The prehumans of Africa approached the threshold of advanced social organization in a manner parallel to that in the lower animals but attained it in a very different manner. As brain size more than doubled, the bands used intelligence based on vastly improved memory. Where primitively social insects evolved division of labor with narrow instincts that play upon categories of social organization in each group, such as larvae and adults, nurses and foragers, the earliest humans operated with variable instinct-driven behavior that made use of detailed knowledge of each group member by all the others.

The creation of groups from personal and intimate mutual knowledge was the unique achievement of humanity. While similarity of genomes by kinship was an inevitable consequence of group formation, kin selection was not the cause. The extreme limitations of kin selection and the phantom-like properties of inclusive fitness apply equally to humans and to eusocial insects and other animals. The origin of the human condition is best explained by the natural selection for social interaction—the inherited propensities to communicate, recognize, evaluate, bond, cooperate, compete, and from all these the deep warm pleasure of belonging to your own special group. Social intelligence enhanced by group selection made *Homo sapiens* the first fully dominant species in Earth's history.

III

OTHER WORLDS

 THE MEANING OF HUMAN EXISTENCE IS BEST
UNDERSTOOD IN PERSPECTIVE, BY COMPARING
OUR SPECIES WITH OTHER CONCEIVABLE LIFE-
FORMS AND, BY DEDUCTION, EVEN THOSE THAT
MIGHT EXIST OUTSIDE THE SOLAR SYSTEM.

7

Humanity Lost in a Pheromone World

Let's continue this journey in a new direction. The greatest contribution that science can make to the humanities is to demonstrate how bizarre we are as a species, and why. The effort to do so is part of research on the natures of all the other species on Earth, each bizarre in its own way. We can go so far as to anticipate a bit the properties of life on other planets, including those that may have evolved human-grade intelligence.

The humanities treat the strange properties of human nature by taking them as "just is." With this perception as a bedrock, creative artists spin stories, make music, and create images in endless detail. The traits that define our species diagnostically appear very narrow when put against the full backdrop of biodiversity. The meaning of human existence cannot be explained until "just is" is replaced with "just is, because."

Let us begin then with the picture of how very specialized and peculiar is our beloved species among the legions of different life-forms that compose Earth's biosphere.

It was only after eons of time, during which millions of species had come and gone, that one of the lineages, the direct antecedents of *Homo sapiens*, won the grand lottery of evolution. The payout was civilization based on symbolic language, and culture, and from these a gargantuan power to extract the nonrenewable resources of the planet—while cheerfully exterminating our fellow species. The winning combination was a randomly acquired mix of preadaptations, which include a life cycle spent entirely on the land, a large brain and the cranial capacity to evolve a still larger brain, free fingers supple enough to manipulate objects, and (this is the hardest part to understand) reliance on sight and sound for orientation instead of smell and taste.

Of course we think ourselves brilliant in our ability to detect chemicals with nose, tongue, and palate. We proudly recognize the bouquet arising from the swirl and aftertaste of a fine vintage. We can identify a darkened room at home from its signature odor alone. Yet we are chemosensory idiots. By comparison most other organisms are geniuses. More than 99 percent of the species of animals, plants, fungi, and microbes rely exclusively or almost exclusively on a selection of chemicals (phero-mones) to communicate with members of the same spe-cies. They also distinguish other chemicals (allomones) to recognize different species of potential prey, preda-tors, and symbiotic partners.

What we enjoy as the sounds of nature are also a small slice of the potential. Bird songs stand out, of course, but bear in mind that birds are among the very few crea-tures that share our dependence for communication on audiovisual channels. Their utterances are joined by the croaking of frogs, chirping of crickets, and shrilling of katydids and cicadas. Add if you wish the twilight chit-tering of bats, although these calls, used for the echolo-cation of obstacles and flying prey, have a pitch above our range of hearing.

Our limited chemosensory skills have profound impli-cations for our relation to the rest of life. So on the side I feel obligated to ask, if flies and scorpions sang as sweetly as songbirds, might we dislike them less?

Turning to the visual signals of animal communica-tion, we enjoy the dances and body coloration of birds, fishes, and butterflies. There are also the brilliant colors and displays of insects, frogs, and snakes used to warn off would-be predators. The messages are urgent and not intended for the delectation of the predators, but instead say, "If you eat me you will die, or get sick, or at least you will hate the taste." Naturalists have a rule about these warnings. If an animal is beautiful and also appears indifferent to your close approach, it is not merely venomous but probably even deadly. Examples include slow-moving coral snakes and insouciant poison-dart frogs. We can see this much, and both enjoy and survive, but we cannot see ultraviolet, with which many insects order their lives—butterflies, for example, in search of ultraviolet-radiating flowers.

The audiovisual signals of the living world excite our emotions and throughout history have often inspired great creative works, the best of music, dance, literature, and the visual arts. They are nevertheless of themselves all paltry compared to what goes on around us in the world of pheromones and allomones. To illustrate this humbling principle of biology, imagine that you had the power to see these chemicals as vividly as the rest of life all around you that smells them.

You are thrust instantly into a world far more dense, complex, and fast-moving than the one you left behind or even imagined. This is the real world of Earth's majority biosphere. Other organisms live in it, but until now you have only lived on the edge. Billowing clouds lift off the ground and vegetation. Odorant tendrils leak out from beneath your feet. Breezes pull all this up past the tops of the trees, where in the stronger winds the tendrils are quickly torn apart and dissipate. Below the ground, confined by litter and soil, wisps arise from rootlets and fungal hyphae, then seep through nearby crevices. The combinations of odors vary from site to site, separated at distances as short as a millimeter away. They form patterns and serve as guideposts—used by ants and other small invertebrates all the time but beyond your meager capacity as a human. In the midst of the background odor field, rare and unusual organic chemicals flow in ellipsoidal streams and expand in hemispherical bubbles. These are the chemical messages emitted by thousands of species of small organisms. Some are produced as effluents evaporating off their bodies. These serve predators as a guide to prey, and equally they serve prey as a warning of approaching predators. Some are messages to others of the same species. "I'm here," they whisper to potential mates and symbiotic partners. "Come, come, please come to me." To potential competitors of the same species, like the pheromones deposited by dogs onto fireplugs, they warn, "You're in my territory. Get out!"

Researchers over the past half century (I had a wonderful time as one of them during the early years, working on communication in ants) have discovered that pheromones are not just broadcast into the air and water for others to pick up. Instead they are aimed with precision at specific targets. The key to understanding any pheromonal communication is the "active space." Whenever odor molecules drift outward from their source (most commonly from a gland in the body of an animal or other organism), a concentration remains within the center of the plume produced that is high enough to be detected by other organisms of the same species. To a remarkable degree, the evolution of each species over thousands or millions of years has engineered the size and structure of the molecule, as well as the amount released in each message, and finally the sensitivity of smell to it in the receiving organism.

Consider a female moth summoning males of her species in the night air. The nearest available male may be a kilometer away—the equivalent of about fifty miles when converted from moth body length to human body length. As a consequence the sex pheromone must be powerful, and so it has proved in real cases studied by pheromone researchers. A male Indian meal moth, for example, is stirred to action by as few as 1.3 million molecules per cubic centimeter. You might think that is a lot of pheromone, but it is actually a vanishingly small amount when compared with, say, a gram of ammonia (NH^3), which contains 10^{23} molecules (one hundred billion trillion) molecules. The pheromone molecule needs to be not only powerful to attract the right kind of male, but also of some rare structure or other, making it highly unlikely to attract a male of the wrong species—or worse, a moth-eating predator. So precise are some sex attractants of moths that those of closely related species differ only by a single atom, or possession or location of a double bond, or even just an isomer.

The male moth of species with such a high level of exclusivity faces a severe problem in finding a mate. The ghostlike active space he must enter and follow starts at a pinpoint on the female's body. It proceeds as a roughly ellipsoid (spindle-shaped) entity until it finally dribbles out to a second pinpoint, then disappears. In most cases the male cannot find the target female by simply moving from a weak concentration of odor along a gradient of ever-increasing concentration, as we do when sniffing out the source of a hidden kitchen smell. It uses another, but at least equally effective method. Upon encountering the pheromone plume the male flies upwind until he reaches the calling female. If he loses the active space, which can happen easily as a breeze shifts and warps the odor stream, he zigzags from side to side through the air until he enters the active space again.

The same magnitude of olfactory power this requires is commonplace throughout the living world. Male rattlesnakes find willing females by following pheromone tracks. Both sexes, their tongues flicking in and out to smell the ground, close in on a chipmunk with no less the precision of a hunter tracking a mallard duck with the barrel of his gun.

The same degree of olfactory skill exists anywhere in the animal kingdom whenever there is a need to make fine discriminations. Among mammals, including human beings, mothers can distinguish the odor of their own infants from those of others. Ants can separate nestmates from aliens in tenths of a second with sweeps of their two antennae over the bodies of approaching workers.

The design of the active space has evolved to communicate many kinds of information in addition to sex and recognition. Guard ants inform nestmates of the approach of enemies by releasing alarm substances. These chemicals are simple in structure compared to sex and trail pheromones. They are released in large quantities, and their active spaces travel far and fast. There is no need for privacy. On the contrary, there is good reason for friend and foe alike to smell them—and the sooner the better. The purpose is to stir alertness and action, and among as many nestmates as possible. Pumped-up fighters rush into the field upon detecting an alarm pheromone, while at the same time nurses carry the young deeper into the nest.

A remarkable pheromone-and-allomone combination is used as a "propaganda substance" by an American species of slave-maker ant. Slavery is widespread in ants of the north temperate zone. It starts when colonies of the slave-making species conduct raids on other ant species. Their workers are shiftless at home, seldom engaging in any domestic chore. However, like indolent Spartan warriors of ancient Greece, they are also ferocious in combat. In some species the raiders are armed with powerful sickle-shaped mandibles capable of piercing the bodies of their opponents. During my research on ant slavery I found one species that uses a radically different method. The raiders carry a hugely enlarged gland reservoir in their abdomen (the rear segment of the three-part body) filled with an alarm substance. Upon breaching the victim's nest, they spray large quantities of the pheromone through the chamber and galleries. The effect on the defenders of the allomone (or, more precisely, pseudo-pheromone) is confusion, panic, and retreat. They suffer the equivalent of our hearing a thunderously loud, persistent alarm coming from all directions. The invaders do not respond the same way. Instead, they are attracted to the pheromone, and as a result they are able easily to seize and carry away the young (in the pupal stage) of the defenders. When the captives emerge from the pupae as adults, they become imprinted, act as sisters of their captors, and serve them willingly as slaves for the rest of their lives.

Ants are possibly the most advanced pheromonal creatures on Earth. They have more olfactory and other sensory receptors on their antennae than any other known kind of insect. They are also walking batteries of exocrine glands, each of which is specialized to produce different kinds of pheromones. In regulating their social lives they employ, according to species, from ten to twenty kinds of pheromones. Each one conveys a different meaning. And that is just the beginning of the information system. Pheromones can be discharged together to create more complex signals. When released at different times or in different places, their meaning changes yet another way. Still more information can be transmitted by varying the concentration of the molecules. In at least one American species of harvester ant I have studied, for example, a barely detectable level of pheromone evokes attention and movement by workers toward the source. A somewhat higher concentration causes the ants to search excitedly back and forth. The highest concentration of the substance, that occurring close to the signaling worker, causes a frenzied attack on any foreign organic object in the vicinity.

Plants of some species communicate by pheromones. At least they are able to read the distress of neighboring plants by responding with actions of their own. A plant attacked by a serious enemy—bacteria, fungus, or insect—releases chemicals that suppress the invader. Some of the substances are volatile. They are "smelled" by their neighbors, who make the same defensive response even though they themselves are not yet under attack. Some species are assaulted by sap-sucking aphids, insects especially abundant in the north temperate zone and capable of wreaking heavy damage. The plant-generated airborne vapor not only stirs neighboring plants to secrete defensive chemicals, but also reaches small wasps that parasitize aphids, drawing them to the vicinity. A few species use yet another defensive line. The signals are transmitted from plant to plant along the strands of symbiotic fungi that entwine the roots and connect one plant to another.

Even bacteria order their lives with pheromone-like communication. Individual cells come together, at which time they trade DNA of special value to one or the other. As their populations increase in density, some species also engage in "quorum sensing." The response is triggered by chemicals released in the liquid around the cells. Quorum sensing results in cooperative behavior and formation of colonies. The best studied of the latter process is the construction of biofilms: free-swimming cells gather, settle on a surface, and secrete a substance that surrounds and protects the entire group. These organized micro-societies are all around and inside us. Among the most familiar are the scum on unwashed bathroom surfaces and the plaque on your inadequately brushed teeth.

There is a simple evolutionary explanation for why our species has taken so long to comprehend the true nature of the pheromone-saturated world in which we live. To start, we are too big to understand the lives of insects and bacteria without special effort. Also, it was necessary while evolving to the *Homo sapiens* level for our forebears to have a large brain, containing memory banks expansible to a size large enough to make possible the origin of language and civilization. Further, bipedal locomotion freed their hands, allowing the construction of increasingly sophisticated tools. Large size and bipedalism together lifted their heads higher than those of any animals other than elephants and a few exceptionally large ungulates. The result was a separation of eyes and ears from almost all the remainder of life. More than 99 percent of the species are too minute in size and bound to the earth far below our senses to receive our ready attention. Finally, our antecedents had to use the audiovisual channel to communicate, not the pheromonal. Any other sensory channel, including pheromones, would have been too slow.

In a nutshell, the evolutionary innovations that made us dominant over the rest of life also left us sensory cripples. It rendered us largely unaware of almost all the life in the biosphere that we have been so heedlessly destroying. That didn't matter very much in early human history, when humans first spread over Earth in the early logarithmic phase of their population growth. Still present in small numbers at that time, they only skimmed energy and resources from the abounding and unsmelled life of the land and sea. There was still enough time and enough room to tolerate a large margin of error. Those happy days have ended. We cannot talk in the language of pheromones, but it will be well to learn more about how other organisms do it, in order better to save them and with them the majority part of the environment on which we depend.

8

The Superorganisms

Imagine yourself a tourist in an East African park, binoculars raised, watching lions, elephants, and a medley of buffalo and antelopes—the iconic large mammals of the savanna. Suddenly one of the continent's greatest and least understood wildlife spectacles of all springs from the ground a few meters in front of you. It is a colony of millions of driver ants emerging from their subterranean nest. They are excited, fast, mindless, a river of small, random furies. At first a teeming mob with no evident purpose, the ants soon form a column that lengthens outward, so densely packed that many of them walk over one another, and the whole comes to resemble a twisting, writhing bundle of ropes.

No living creature dares to touch the angry column. Every one of the foragers is ready to bite and sting furiously any intrusive object that might serve as food. Posted along the column are soldiers, big defense specialists that stand on raised legs with pincer-shaped mandibles poised upward. The driver ants are well organized, yet they have no leaders. The vanguard consists of whichever of the blind workers happen to reach the front at the moment. These dash forward briefly before yielding to others that press from behind.

At twenty meters or so out from the nest, the point of the column begins to spread into a fan-shaped front, composed of smaller and then still smaller columns. Very quickly the ground in its path is covered with a network of columns and individual workers, hunting and seizing insects, spiders, and other invertebrates. The purpose of the foray now becomes apparent. The ants are universal predators, harvesting as many small prey as they can subdue and bring back to the nest as food. The columns also drag home entire or in pieces any larger animals unable to get out of their way—lizards, snakes, small mammals, and, it is rumored, occasional unguarded babies. There is good reason for the unrelenting ferocity of the driver ants. A multitude of mouths must be fed with a lot of food, and frequently, because if not, the whole system will soon collapse. The entire colony, foragers and home-bound workers combined, consists of as many as twenty million sterile females. All are daughters of the thumb-sized mother queen, which not surprisingly is also the largest ant known in the world.

The driver ant colony is one of the most extreme superorganisms ever evolved. If as you watch it you blur your focus a little, it resembles a gigantic amoeba sending out a meters-long pseudopodium to engulf particles of food. The units of the superorganism are not cells, as they are in amoebas and other organisms, but individual, full-bodied, six-legged organisms. These ants, these organism-units, are totally altruistic to one another and coordinated so completely that they closely resemble the combined cells and tissue of an organism. When you see them in nature or on film, you cannot help describing the driver ant colony as "it" rather than "they."

All of the fourteen thousand known species of ants form colonies that are superorganisms, although only a very few are as complexly organized or as large as the driver ants. For nearly seven decades, starting in boyhood, I've studied hundreds of kinds of ants around the world, both simple and complex. This experience qualifies me, I believe, to offer you some advice on ways their lives can be applied to your own life (but, as you will see, of very limited practical use). I'll start with the most frequent question I'm asked by the general public: "What can I do about the ants in my kitchen?" My response comes from the heart: Watch your step, be careful of little lives. They especially like honey, tuna, and cookie crumbs. So put down bits of those on the floor, and watch closely from the moment the first scout finds the bait and reports back to her colony by laying an odor trail. As a little column follows her out to the food, you will see social behavior so strange it might be on another planet. Think of the kitchen ants not as pests or bugs, but as your personal guest superorganisms.

The second most frequently asked question is, "What can we learn of moral value from the ants?" Here again I will answer definitively. Nothing. Nothing at all can be learned from ants that our species should even consider imitating. For one thing, all working ants are female. Males are bred and appear in the nest only once a year, and then only briefly. They are unappealing, pitiful creatures with wings, huge eyes, small brain, and genitalia that make up a large portion of their rear body segment. They do no work while in the nest and have only one function in life: to inseminate the virgin queens during the nuptial season when all fly out to mate. They are built for their one superorganismic role only: robot flying sexual missiles. Upon mating or doing their best to mate (it is often a big fight for a male just to get to a virgin queen), they are not admitted back home, but instead are programmed to die within hours, usually as victims of predators. Now for the moral lesson: although like almost all well-educated Americans I am a devoted promoter of gender equality, I consider sex practiced the ant way a bit extreme.

To return, briefly, to life in the nest, many kinds of ants eat their dead. Of course that's bad enough—but I'm obliged to tell you they also eat their injured. You may have seen ant workers retrieve nestmates that you have mangled or killed underfoot (accidentally, not deliberately, I hope), thinking it battlefield heroism. The purpose, alas, is more sinister.

As ants grow older, they spend more time in the outermost chambers and tunnels of the nest, and are more prone to undertake dangerous foraging trips. They also are the first to attack enemy ants and other intruders that swarm into their territories and around their nest entrances. Here indeed is a major difference between people and ants: where we send our young men to war, ants send their old ladies. No moral lesson there, unless you are looking for a less expensive form of elder care.

Ants that are ill move along with the aged to the nest perimeter, and even to the outside. There being no ant doctors, leaving home is not to find an ant clinic but solely to protect the rest of the colony from contagious disease. Some ants die of fungus and trematode worm infections outside the nest, allowing these organisms to disseminate their own offspring. This behavior can be easily misinterpreted. You may wonder, if you have seen too many Hollywood films on alien invaders and zombies, as I have, whether the parasite is controlling the brain of its host. The reality is much simpler. The sick ant has a hereditary tendency to protect its nestmates by leaving the nest. The parasite, for its part, has evolved to take advantage of ants that are socially responsible.

The most complex societies of all ant species, and arguably of all animals everywhere, are the leafcutters of the American tropics. In lowland forests and grasslands from Mexico to warm temperate South America, you find conspicuous long files of reddish, medium-sized ants. Many carry freshly cut pieces of leaves, flowers, and twigs. The ants drink sap but don't eat solid fresh vegetation. Instead, they carry the material deep into their nests, where they convert it into numerous complex, spongelike structures. On this substrate they grow a fungus, which they do eat. The entire process, from collection of raw plant material to the final product, is conducted in an assembly line employing a sequence of specialists. The leafcutters in the field are medium in size. As they head home with their burdens, unable to defend themselves, they are harassed by parasitic phorid flies eager to deposit eggs that hatch into flesh-eating maggots. The problem is solved, mostly, by tiny sister ant workers that ride on their backs, like mahouts on elephants, and chase the flies away with flicks of their hind legs. Inside the nest, workers somewhat smaller than the gatherers scissor the fragments into pieces about a millimeter in diameter. Still smaller ants chew the fragments into lumps and add their own fecal material as fertilizer. Even smaller workers use the gooey lumps thus created to construct the gardens. The smallest workers—the same size as the anti-fly guards—plant and tend the fungus in the gardens.

There is one additional caste of leafcutter ants, comprising the largest workers of all. They have outsized heads swollen with adductor muscles, which close their razor-sharp mandibles with enough force to slice leather (not to mention your skin). They appear to be specialized to defend against the most dangerous predators, including especially anteaters and probably a few other sizable mammals. The soldiers stay hidden deep in the lower chambers, charging forth only when the nest is in serious trouble. During a recent field trip in Colombia, I stumbled on a way to bring these brutes to the surface with almost no effort. I knew that leafcutter nests are constructed as a giant air-conditioning system. Channels near the center accumulate exhausted, CO_2-laden air heated by the gardens and the millions of ants living on them. As the air is warmed, it moves by convection through openings directly above. At the same time fresh air is pulled into the nest through openings to channels located around the periphery of the nest. I found that if I blew into the peripheral channels, allowing my mammalian breath to be carried down into the nest center, the big-headed soldiers soon came out looking for me. I admit that this observation has no practical use, unless you like the thrill of being chased by really serious ants.

The advanced superorganisms of ants, bees, wasps, and termites have achieved something resembling civilizations almost purely on the basis of instinct. They have done so with brains one-millionth the size of human brains. And they have accomplished the feat with a remarkably small number of instincts. Think of evolving a superorganism as constructing a Tinkertoy. With just a few basic pieces fitted together in different ways it is possible to manufacture a wide variety of structures. In the evolution of superorganisms, those that survive and reproduce the most effectively are the ones that dazzle us today with their sophisticated complexity.

The fortunate few species able to evolve superorganismic colonies have as a whole also been enormously successful. The twenty thousand or so known species of social insects (ants, termites, social bees, and wasps combined) make up only 2 percent of the approximately one million known species of insects, but three-fourths of the insect biomass.

With complexity, however, comes vulnerability, and

that brings me to one of the other superorganism super-stars, the domestic honeybee, and a moral lesson. When disease strikes solitary or weakly social animals that we have embraced in symbiosis, such as chickens, pigs, and dogs, their lives are simple enough for veterinarians to diagnose and fix most of the problems. Honeybees, on the other hand, have by far the most complex lives of all our domestic partners. There are a great many more twists and turns in their adaptation to their environment that upon failing could damage some part or other of the colony life cycle. The intractability thus far of the honeybee colony collapse disorder of Europe and North America, which threatens so much of crop pollination and humanity's food supply at the present time, may represent an intrinsic weakness of superorganisms in general. Perhaps, like us, with our complex cities and interconnected high technology, it is their excellence that has put them at greater risk.

You may occasionally hear human societies described as superorganisms. This is a bit of a stretch. It is true that we form societies dependent on cooperation, labor specialization, and frequent acts of altruism. But where social insects are ruled almost entirely by instinct, we base labor division on transmission of culture. Also we, unlike social insects, are too selfish to behave like cells in an organism. Almost all human beings seek their own destiny. They want to reproduce themselves, or at least enjoy some form of sexual practice adapted to that end. They will always revolt against slavery; they will not be treated like worker ants.

9

Why Microbes Rule the Galaxy

Beyond the Solar System there is life of some kind. It exists, experts agree, on at least a small minority of Earthlike planets that circle stars as close as a hundred light-years to the Sun. Direct evidence of its presence, whether positive or negative, may come soon, perhaps within a decade or two. It will be obtained by spectrometry of light from mother stars that passes through the atmospheres of the planets. If the analysts detect "biosignature" gas molecules of a kind that can be generated only by organisms (or else are far more abundant than expected in a nonliving equilibrium of gases), the existence of alien life will pass from the well-reasoned hypothetical to the very probable.

As a student of biodiversity and, perhaps more importantly, at heart a congenital optimist, I believe I can add credibility to the search for extrasolar life from the history of Earth itself. Life arose here quickly when conditions became favorable. Our planet was born about 4.54 billion years ago. Microbes appeared soon after the surface became even marginally habitable, within one hundred million to two hundred million years. The interval between habitable and inhabited may seem an eternity to the human mind, but it is scarcely a night and a day in the nearly fourteen-billion-year history of the Milky Way galaxy as a whole.

Granted that the origin of life on Earth is only one datum in a very big Universe. But astrobiologists, using an increasingly sophisticated technology focused on the search for alien life, believe that at least a few and probably a large number of planets in our sector of the galaxy have had similar biological geneses. The conditions they seek are that the planets have water and are in "Goldilocks" orbit—not close enough to the mother star to be furnace-blasted, yet not so far away that their water is forever locked in ice. It should also be kept in mind, however, that just because a planet is inhospitable now does not mean it always has been so. Further, on a seemingly otherwise barren surface there may exist small pockets of habitats—oases—that support organisms. Finally, life might have originated somewhere with molecular elements different from those in DNA and energy sources used by organisms on Earth.

One prediction seems unavoidable: whatever the condition of alien life, and whether it flourishes on land and sea or barely hangs on in tiny oases, it will consist largely or entirely of microbes. On Earth these organisms, the vast majority too small to see with unaided vision, include most protists (such as amoebae and paramecia), microscopic fungi and algae, and, smallest of all, the bacteria, archaeans (similar in appearance to bacteria but genetically very different), picozoans (ultrasmall protists only recently distinguished by biologists), and viruses. To give you a sense of size, think of one of your own trillions of human cells, or a solitary amoeba or single-celled alga, as the size of a small city. Then a typical bacterium or archaean would be the size of a football field and a virus the size of a football.

Earth's overall microbial fauna and flora are resilient in the extreme, occupying habitats that might at first seem death traps. An extraterrestrial astronomer scanning Earth would not see, for example, the bacteria that thrive in volcanic spumes of the deep sea above the temperature of boiling water, or other bacterial species in mine outflows with a pH close to that of sulfuric acid. The E.T. would not be able to detect the abundant microscopic organisms on the Mars-like surface of Antarctica's McMurdo Dry Valleys, considered Earth's most inhospitable land environment outside of the polar ice caps. E.T. would be unaware of *Deinococcus radiodurans*, an

Earth bacterium so resistant to lethal radiation that the plastic container in which it is cultured discolors and cracks before the last cell dies.

Might other planets of the Solar System harbor such extremophiles, as we Earth biologists call them? On Mars, life could have evolved in the early seas and survive today in deep aquifers of liquid water. Abundant parallels of such subterranean regression exist on Earth. Advanced cave ecosystems abound on all continents. They include at the least microbes, and in most parts of the world insects and spiders and even fish as well, all with anatomy and behavior specialized for life in totally dark, impoverished environments. Even more impressive are the SLIMEs (subterranean lithoautotrophic microbial ecosystems), distributed through soil and rock fissures from near the surface to a depth of up to 1.4 kilometers and comprising bacteria that live on energy drawn from the metabolism of rocks. Feeding on them are a recently discovered new species of deep subterranean nematodes, tiny worms of a general kind abundant everywhere on the surface of the planet.

There are places in the Solar System in addition to Mars to search for organisms, at least those with the biology of what we call extremophile on Earth. It makes sense to look for microbes in aquatic islets beneath or around the icy geysers of Enceladus, Saturn's superactive little moon. And as opportunity arises, we should (in my opinion) probe the vast aquatic oceans of Jupiter's moons Callisto, Europa, and Ganymede, as well as Titan, a larger moon of Saturn. All are encased in thick shells of ice. Brutally cold and lifeless on the surface they may be, but underneath are depths warm enough to hold liquid organisms. We can eventually, if we wish, drill through the shells to reach that water—just as scientific explorers are now doing above Lake Vostok, sealed off by the Antarctic ice cap for a million years or more.

Someday, perhaps in this century, we, or much more likely our robots, will visit these places in search of life. We must go and we will go, I believe, because the collective human mind shrivels without frontiers. The longing for odysseys and faraway adventure is in our genes.

The ultimate destiny of the outward-bound astronomers and biologists is of course to reach still farther, very much farther, across almost incomprehensibly far distances into space, to the stars and potentially life-bearing planets around them. Because deep space is transparent to light, the detection of very remote alien life is very much a possible dream. Many potential targets will be found in the mass of data collected by the Kepler space telescope before it partly failed in 2013, together with other space telescopes planned and the most powerful ground-based telescopes. And soon. By mid-2013 almost 900 extrasolar planets had been detected, with thousands more believed likely to be found in the near future. One recent extrapolation (let me pause: extrapolation is admittedly a risky procedure in science) predicts that a fifth of stars are orbited by Earth-sized planets. In fact, the most common class of systems detected thus far include planets one to three times Earth's size, thus with gravity similar to that of Earth. So, what does that tell us about the potential of life in outer space? First, consider the estimate that ten stars of various kinds exist within ten light-years of the Sun, about 15,000 within 100 light-years, and 260,000 within 250 light-years. Keeping in mind life's early origin in the geological history of Earth as a clue, it is plausible that the total number of life-bearing planets as close as 100 light-years could be in the tens or hundreds.

To find even the simplest form of extraterrestrial life would be a quantal leap in human history. In self-image, it would confirm humanity's place in the Universe as both infinitely humble in structure and infinitely majestic in achievement.

Scientists will want (desperately) to read the genetic code of the extraterrestrial microbes, providing such organisms can be located elsewhere in the Solar System and their molecular genetics studied. This step is feasible with robotic instruments, eliminating the need of bringing the organisms to Earth. It would reveal which of two opposing conjectures about the code of life is correct. First, if the microbial E.T.s have a code different from that on Earth, their molecular biology would be different to a comparable degree. And if such proves to be the case, an entirely new biology might be instantly created. We would further be forced to conclude that the code used by life on Earth is probably only one of many possible in the galaxy, and that codes in other star systems have originated as adaptations to environments very different from those on Earth. If, on the other hand, the code of extraterrestrials is basically the same as that of native Earth organisms, it could suggest (but not prove, not yet) that life everywhere can only originate with one code, the same as in Earth's biological genesis.

Alternatively, perhaps some organisms manage interplanetary travel by drifting through space, living in cryogenic dormancy for thousands or millions of years, protected somehow from galactic cosmic radiation and surges of solar energetic particles. Interplanetary or even interstellar travel by microbes, called pangenesis, sounds like science fiction. I wince a bit just bringing it up. But it should be considered as at least a remote possibility. We know too little about the vast array of bacteria, archaeans, and viruses on Earth to make any call about the extremes of evolutionary adaptation, here and elsewhere in the Solar System. In fact, we now know that some Earth bacteria are poised to be space travelers, even if (perhaps) none has ever succeeded. A large number of living bacteria occur in the middle and upper atmosphere, at altitudes of six to ten kilometers. Composing an average of around 20 percent of the particles with diameters of 0.25 to 1 micron in diameter, they include species able to metabolize carbon compounds of kinds found all around them in the same strata. Whether some are also able to maintain reproducing populations, or on the contrary are just temporary voyagers lifted by air currents away from the land surface, remains to be learned.

Perhaps the time has come to seine-haul for microbes at varying distances beyond Earth's atmosphere. The nets could be composed of ultrafine sheets towed by orbiting satellites through billions of cubic kilometers of space, then folded and returned for study. Such a foray conducted out into space might produce surprising results. Even new, anomalous species of Earth-born bacteria able to endure the most hostile conditions—or the absence of such organisms—would make the effort worthwhile. It would help answer two of the key questions of astrobiology: What are the extreme environmental conditions in which current members of Earth's biosphere can exist? And might organisms originate in other worlds in conditions of comparable severity?

10

A Portrait of E.T.

What I am about to tell you is speculation, but not pure speculation. It is that by examining the myriad animal species on Earth and their geological history, then extending this information to plausible equivalents on other planets, we can make a rough sketch of the appearance and behavior of intelligent extraterrestrial organisms. Please don't leave me at this point. Refrain from dismissing this approach out of hand. Instead, call it a scientific game, with the rules changing to fit new evidence. The game is well worth playing. The payoff, even if the chance of contact with human-grade aliens or higher proves forever vanishingly small, is the building of a context within which a sharper image of our own species can be drawn.

Granted there is temptation to leave the subject to Hollywood, to the creation of the nightmarish monsters of *Star Wars* or the Americans-in-punk-makeup populating *Star Trek*. Learning about extraterrestrial microbes is one thing: it is not difficult to imagine in broad principles the self-assembly of primitive organisms at the level of Earth's bacteria, archaeans, picozoans, and viruses; and scientists may soon find evidence of such microbial life on other planets. But it is an entirely different matter to picture the origin of extraterrestrial intelligence at the human grade or higher. This most complex level of evolution has occurred on Earth only once, and then only after more than six hundred million years of evolution within a vast diversity of animal life.

The final evolutionary steps prior to the human-level singularity, that is, altruistic division of labor at a protected nest site, has occurred on only twenty known occasions in the history of life. Three of the lines that reached this final preliminary level are mammals, namely two species of African mole rats and *Homo sapiens*—the latter a strange offshoot of African apes. Fourteen of the twenty high achievers in social organization are insects. Three are coral-dwelling marine shrimp. None of the nonhuman animals has a large enough body, and hence potential brain size, needed to evolve high intelligence.

That the prehuman line made it all the way to *Homo sapiens* was the result of our unique opportunity combined with extraordinarily good luck. The odds opposing it were immense. Had any one of the populations directly on the path to the modern species suffered extinction during the past six million years since the human-chimpanzee split—always a dire possibility, since the average geological life span of a mammal species is about five hundred thousand years—another *hundred million* years might have been required for a second human-level species to appear.

Because of all of the pieces that likely must also fall in place beyond the Solar System, intelligent E.T.s are also likely to be both improbable and rare. Given that, and assuming they exist at all, it is reasonable to ask how close to Earth might E.T.s at the human grade or higher be found. Allow me an educated guess. Consider first the many thousands of large terrestrial animal species that have flourished on Earth for the past four hundred million years, with none but our own making the ascent. Next, consider that while 20 percent or more star systems may be circled by Earthlike planets, only a small fraction may carry liquid water and also possess a Goldilocks orbit (to remind you, not so close to the mother star to be baked, not so distant to be kept permanently deep-frozen). These pieces of evidence are admittedly very slender, but they give reason to doubt that high intelligence has evolved in any of the 10 star systems within 10 light-years of the Sun. There is a chance, slight but otherwise impossible to judge reliably, that the event has occurred within a distance of 100 light-years of the Sun, a radius encompassing 15,000 star systems. Within 250 light-years (260,000 star systems), the odds are dramatically increased. At this distance, if we work strictly off the experience of Earth, the uncertain and marginally possible changes to the probable.

Let's grant the dream of many science fiction writers and astronomers alike that civilized E.T.s are out there, even if at this almost incomprehensible distance. What might they be like? Permit me to make a second educated guess. By combining the evolution and peculiar properties of hereditary human nature with known adaptations by millions of other species in the great biodiversity of Earth, I believe it's possible to produce a logical albeit very crude hypothetical portrait of human-grade aliens on Earth-like planets.

E.T.s are fundamentally land-dwellers, not aquatic. During their final ascent in biological evolution to the human grade of intelligence and civilization, they must have used controlled fire or some other easily transportable high-energy source to develop technology beyond the earliest stages.

E.T.s are relatively large animals. Judging from Earth's most intelligent terrestrial animals—they are, in descending rank order, Old World monkeys and apes, elephants, pigs, and dogs—E.T.s on planets with the same mass as Earth or close to it evolved from ancestors that weighed between ten and a hundred kilograms. Smaller body size among species means smaller brains on average, along with less memory storage capacity and lower intelligence. Only big animals can carry on board enough neural tissue to be smart.

E.T.s are biologically audiovisual. Their advanced technology, like our own, allows them to exchange information at various frequencies across a very broad sector of the electromagnetic spectrum. But in ordinary thinking and talking among themselves they use vision just like us, employing a narrow section of the spectrum, along with sound created with waves of air pressure. Both are needed for rapid communication. E.T.s' unaided vision may allow them to see the world in ultraviolet in the manner of butterflies, or some other, still unnamed primary color outside the range of wave frequency sensed by humans. Their auditory communication may be immediately perceived by us, but it could also easily be at too high a pitch, as used by katydids or many other insects, or too low, as practiced by elephants. In the microbial worlds on which the E.T.s depend, and in probably most of the animal world, most communication is by pheromones, secreted chemicals that convey meaning in their smell and taste. The E.T.s, however, cannot employ this medium any more than we can. While it is theoretically possible to send complex messages by the controlled release of odor, the frequency and amplitude modulation required to create a language is possible across only a few millimeters.

Finally, might E.T.s read facial expressions or sign language? Of course. Thought waves? Sorry, I don't see any way that's possible, except through elaborate neurobiological technology.

Their head is distinct, big, and located up front. The bodies of all land-dwelling animals on Earth are elongated to some extent, and most are bilaterally symmetrical, with the left and right sides of their bodies reciprocal mirror images. All have brains with key sensory input located in the head, adapted in location for quick scanning, and integration, and action. E.T.s are no different. The head is also large compared to the rest of the body, with a special chamber to accommodate the necessarily huge memory banks.

They possess light to moderate jaws and teeth. Heavy mandibles and massive grinding teeth on Earth are the marks of dependence on coarse vegetation. Fangs and horns denote either defense against predators, or competition among males of the same species, or both. During their evolutionary ascent, the ancestors of the aliens almost certainly relied on cooperation and strategy rather than brute strength and combat. They were also likely to be omnivorous, as are humans. Only a broad, high-energy meat-and-vegetable diet could produce the relatively large populations needed for the final stage of the ascent—which in humans occurred with the invention of agriculture, villages, and other accoutrements of the Neolithic revolution.

They have very high social intelligence. All social insects (ants, bees, wasps, termites) and the most intelligent mammals live in groups whose members continuously and simultaneously compete and cooperate with one another. The ability to fit into a complex and fast-moving social network gives a Darwinian advantage both to the groups and to individual members that form them.

E.T.s have a small number of free locomotory appendages, levered for maximum strength with stiff internal or external skeletons composed of hinged segments (as by human elbows and knees), and with at least one pair of which are terminated by digits with pulpy tips used for sensitive touch and grasping. Since the first lobe-finned fishes invaded the land on Earth about four hundred million years ago, all of their descendants, from frogs and salamanders to birds and mammals, have possessed four limbs. Further, among the most successful and abundant land-dwelling invertebrates are the insects, with six locomotory appendages, and spiders, with eight. A small number of appendages is therefore evidently good. It is moreover the case that only chimpanzees and humans invent artifacts, which vary in nature and design from one culture to the next. They do so because of the versatility of soft fingertips. It is hard to imagine any civilization built with beaks, talons, and scrapers.

They are moral. Cooperation among group members based on some amount of self-sacrifice is the rule among highly social species on Earth. It has arisen from natural selection at both the individual and group levels, and especially the latter. Would E.T.s have a similar inborn moral propensity? And would they extend it to other forms of life, as we have done (however imperfectly) in biodiversity conservation? If the driving force of their early evolution is similar to our own, a likely possibility, I believe they would possess comparable moral codes based upon instinct.

It might not have escaped your attention that I've thus far tried to envision E.T.s only as they were at the beginning of their civilizations. It is the equivalent of a portrait of humanity drawn during the Neolithic era. Following that period our species worked its way by cultural evolution, across ten millennia, from the rudiments of civilization in scattered villages to the technoscientific global community of today. It is likely by chance alone that extraterrestrial civilizations made the same leap not just millennia ago but thousands of millennia ago. With the same intellectual capacity we already have, and possibly a great deal more, might they have long since engineered their own genetic code in order to change their biology? Did they enlarge their personal memory capacities and develop new emotions while diminishing old ones—thereby adding boundless new creativity to the sciences and the arts?

I think not. Nor will humans, other than correcting disease-causing mutant genes. I believe it would be unnecessary for our species' survival to retrofit the human brain and sensory system, and, in one basic sense at least, it would be suicidal. After we have made all of the cultural knowledge available with only a few keystrokes, and after we have built robots that can outthink and outperform us, both of which initiatives are already well under way, what will be left to humanity? There is only one answer: we will choose to retain the uniquely messy, self-contradictory, internally conflicted, endlessly creative human mind that exists today. That is the true Creation, the gift given us before we even recognized it as such or knew its meaning, before movable print and space travel. We will be existential conservatives, choosing not to invent a new kind of mind grafted on top of or supplanting the admittedly weak and erratic dreams of our old mind. And I find it comforting to believe that smart E.T.s, wherever they are, will have reasoned the same way.

Finally, if E.T.s know of Earth's existence at all, will they choose to colonize it? In theory, it may have seemed possible and contemplated at any time by many of them over the past millions or hundreds of millions of years. Suppose a conqueror E.T. species has arisen somewhere in our neighborhood of the galaxy since the time of Earth's Paleozoic Era. Like our species, it was from the beginning driven by an impulse to invade all of the habitable worlds it could reach. Imagine that its drive for cosmic lebensraum began one hundred million years ago, in an already old galaxy. Also, imagine (reasonably) that it took ten millennia from launch to reach the first habitable planet; and from there, with the technology perfected, the colonists devoted another ten millennia to launch an armada sufficient to occupy ten more planets. By continuing this exponential growth, the hegemons would have already colonized most of the galaxy.

I'll give you two good reasons why galactic conquests have never happened, or even begun, and hence why our poor little planet has not been colonized and never will be. A remote possibility exists that Earth has been visited by sterile robot probes, or in some distant future age might yet be visited, but they will not be accompanied by their organic creators. All E.T.s have a fatal weakness. Their bodies would almost certainly carry microbiomes, entire ecosystems of symbiotic microorganisms comparable to the ones that our own bodies require for day-to-day existence. The E.T. colonists would also be forced to bring crop plants, algae-equivalents, or some other energy-gathering organisms, or at the very least synthetic organisms to provide their food. They would correctly assume that every native species of animal, plant, fungus, and microorganism on Earth is potentially deadly to them and to their symbionts. The reason is that the two living worlds, ours and theirs, are radically different in origin, molecular machinery, and the endless pathways of evolution that produced the life-forms then brought together by colonization. The ecosystems and species of the alien world would be wholly incompatible with our own.

The result would be a biological train wreck. The first to perish would be the alien colonists. The residents—us and all of Earth's fauna and flora, to which we are so exquisitely well adapted—would be unaffected except briefly and very locally. The clash of worlds would not be the same as the ongoing exchange of species of plants and animals between Australia and Africa, or between North and South America. It's true that considerable damage to native ecosystems has recently occurred due to such intercontinental mixing, caused by our own species. Many of the colonists hang on as invasive species, especially in habitats disturbed by humans. A few manage to crowd native species to extinction. But it is nothing like the vicious biological incompatibility that would doom interplanetary colonists. In order to colonize a habitable planet, the aliens would first have to destroy all life on it, down to the last microbe. Better to stay at home, for a few more billion years anyway.

This brings me to the second reason why our fragile little planet has nothing to fear from extraterrestrials. E.T.s bright enough to explore space surely also understand the savagery and lethal risk inherent in biological colonization. They would have come to the realization, as we have not, that in order to avoid extinction or reversion to unbearably harsh conditions on their home planet they had to achieve sustainability and stable political systems long before journeying beyond their star system. They may have chosen to explore other life-bearing planets—very discreetly with robots—but not to undertake invasions. They had no need, unless their home planet was about to be destroyed. If they had developed the ability to travel between star systems, they would also have developed the ability to avoid planetary destruction.

There live among us today space enthusiasts who

believe humanity can emigrate to another planet after using up this one. They should heed what I believe is a universal principle, for us and for all E.T.s: there exists only one habitable planet, and hence only one chance at immortality for the species.

The Collapse of Biodiversity

Think of Earth's biodiversity, the planet's variety of life, as a dilemma wrapped in a paradox. The paradox is the following contradiction: the more species that humanity extinguishes, the more new species scientists discover. However, like the conquistadores who melted the Inca gold, they recognize that the great treasure must come to an end—and soon. That understanding creates the dilemma: whether to stop the destruction for the sake of future generations, or the opposite, just go on changing the planet to our immediate needs. If the latter, planet Earth will recklessly and irreversibly enter a new era of its history, called by some the Anthropocene, an age of, for, and all about our one species alone, with all the rest of life rendered subsidiary. I prefer to call this miserable future the Eremocene, the Age of Loneliness.

Scientists divide biodiversity (keep in mind, I mean all the rest of life) into three levels. At the top are the ecosystems, for example meadows, lakes, and coral reefs. Below it are the species that make up each of the ecosystems in turn. And at the base are the genes that prescribe the distinguishing traits of each of the species.

A convenient measure of biodiversity is the number of species. When in 1758 Carl Linnaeus began the formal taxonomic classification still in use today, he recognized about twenty thousand species in the entire world. He thought that he and his students and helpers might be able to account for most or all of the world fauna and flora. By 2009, according to the Australian Biological Resources Study, the number had grown to 1.9 million. By 2013 it was probably 2 million. Yet this is still only an early point in the Linnaean journey. The actual number in nature is not known even to the nearest order of magnitude. When still-undiscovered invertebrates, fungi, and microorganisms are added, estimates vary wildly, from five million to one hundred million species.

Earth, to put the matter succinctly, is a little-known planet. The pace of mapping biodiversity has also remained slow. New species flood laboratories and museums everywhere, but are being diagnosed and named at a pace of only about twenty thousand a year. (I have described about 450 new species of ants from around the world during my lifetime.) At this rate, and taking a low-end estimate of five million species remaining to be classified, the task will not be completed until the middle of the twenty-third century. Such a snail's pace is a disgrace of the biological sciences. It is based on the misconception that taxonomy is a completed and outdated part of biology. As a result this still-vital discipline has been largely squeezed out of academia and relegated to natural history museums, themselves impoverished and forced to reduce their research programs.

The exploration of biodiversity has few friends in the corporate and medical world. This is a serious mistake. Science as a whole loses as a result. Taxonomists do far more than name species. They are also experts and primary researchers on the organisms of their specialty. To them we must turn for most of what is known on nonhuman life, including world-dominant groups such as nematodes, mites, insects, spiders, copepods, algae, grasses, and composites on which our own lives ultimately depend.

The fauna and flora of an ecosystem are also far more than collections of species. They are also a complex system of interactions, where the extinction of any species under certain conditions could have a profound impact on the whole. It is an inconvenient truth of the environmental sciences that no ecosystem under human pressure can be made sustainable indefinitely without knowing all of the species that compose it, which commonly number in the thousands or more. The knowledge coming from taxonomy and biological studies dependent upon it are as necessary for ecology as are anatomy and physiology for medicine.

Otherwise, scientists easily misjudge which ones are likely to be "keystone" species—those on which the life of the ecosystem is dependent. The most potent keystone species demonstrated in the world may be the sea otter, a cat-sized cousin of weasels that lives along the coast from Alaska to southern California. Because its luxurious fur was so much prized, the species was hunted to near-extinction by the end of the nineteenth century—with a catastrophic ecological result. The kelp forest, a dense mass of algal vegetation anchored on the sea bottom, reaching to the surface, home to a vast number of shallow marine species, and nursery for other species from deeper waters, also mostly disappeared. The cause: sea otters feed heavily on sea urchins, and these spiny invertebrates feed heavily on kelp. When the sea otters were taken out, the sea urchin population exploded, and large sections of the ocean floor were reduced to desertlike surfaces called sea-urchin barrens. When the sea otter populations were protected and allowed to flourish again, the sea urchins declined and the kelp forests returned.

How can we care for species composing Earth's living environment if we don't even know the great majority of them? Conservation biologists agree that large numbers of species are going extinct before they are discovered. Even in purely economic terms, the opportunity costs of extinction are going to prove enormous. Research on just small numbers of wild species has yielded major advances in the quality of human life—an abundance of pharmaceuticals, new biotechnology, and advances in agriculture. If there were no fungi of the right kind, there would be no antibiotics. Without wild plants with edible stems, fruit, and seeds available for selective breeding, there would be no cities, and no civilization. No wolves, no dogs. No wild fowl, no chickens. No horses and camelids, no overland journeys except by hand-pulled vehicles and backpacks. No forests to purify water and pay it out gradually, no agriculture except with less productive dryland crops. No wild vegetation and phytoplankton, not enough air to breathe. Without nature, finally, no people.

The human impact on biodiversity, to put the matter as briefly as possible, is an attack on ourselves. It is the action of a mindless juggernaut fueled by the biomass of the very life it destroys. The agents of destruction are summarized by the acronym HIPPO, with the relative importance of the agents declining left to right, in this acronym, in most parts of the world:

Habitat loss (H), by far the leading agent of destruction, is defined as the reduction of habitable area by deforestation, conversion of grassland, and that great golem arising from all our excesses, climate change.

Invasive species (I), aliens that cause damage to humans or the environment or both, create global havoc. Their variety and number in every country for which counts have been made is increasing exponentially. Despite improving quarantines, the immigrants pour in faster and faster. South Florida now has a varied fauna of parrots where none existed before (except for the now-extinct Carolina parakeet), and two species of pythons, one each from Asia and Africa, that compete with American alligators at the top of the food chain.

Hawaii is the American capital of extinction, having lost more of its endemic plants, birds, and insects— those species and subspecies found nowhere else—by a wider margin than any other state. Its endemic species of birds are down to forty-two from the original seventy-one estimated present when the first Polynesians came ashore over a thousand years ago. They have been hammered at two levels. The accidental introduction of mosquitoes in the nineteenth century allowed the spread of avian pox. Feral pigs, while rooting through the soil of the upland forests, churn the ground into foul muck and silt, which eventually helps create long-lasting pools of water ideal for mosquito larvae.

Equally deadly on a global scale has been the human-aided transport of the chytrid fungus *Batrachochytrium dendrobatidis*, a parasite of frogs, into the American tropics and Africa. The parasite evidently travels in aquaria containing infected animals. The fungus spreads through the skin and, because frogs breathe through their skins, suffocates its host. Scores of frog species have been extinguished or threatened with extinction.

And if this were not enough, there are invasive plant species capable of destroying a complete ecosystem. Such is the velvet tree *Miconia calvescens*, a beautiful small tree of the American tropics grown widely around the world as an ornamental. On the islands of Polynesia it has also proved to be a menace capable, when not controlled, of growing to full size and in stands so dense as to crowd out all other plant species and most forms of animal life as well.

Pollution (first P in the HIPPO series) has inflicted most of its damage to fish and other life in freshwater systems. But it is also the cause of the more than four hundred anoxic "dead zones" in marine waters that receive contaminated water from upstream agricultural land.

Population growth (the second P) is actually a catalytic force of all the other factors. Damage will not be so much from the growth itself, which is expected to peak by the end of the century, but rather from the rapid and unstoppable ascent in per capita consumption worldwide as economies improve.

Finally, the role of overharvesting (O) is best illustrated by the percentage of global decline in the catch of various species of marine pelagic fishes such as tuna and swordfish from the mid-1850s to the present: 96 to 99 percent. Not only are these species scarcer, but the individual fish caught are on average also smaller.

Of course there exists an earnest worldwide effort to map and save biodiversity. The Census of Marine Life and Encyclopedia of Life programs have made available on the internet most of what we know of Earth's species. New techniques are helping to discover new species and to identify those already named, with increased speed and precision. Most notable among these methods is barcoding, the identification of species by reading short sections of highly variable DNA. Global conservation organizations such as Conservation International, World Wildlife Fund–U.S., and the International Union for Conservation of Nature, along with a host of governmental and private organizations, are doing all they can—often with heroic exertions—to stem the hemorrhaging of biodiversity.

How much has this effort achieved? In 2010 a team of experts drawn from 155 research groups around the world joined to assess the status of 25,780 vertebrate species (mammals, birds, reptiles, amphibians, and fishes). These were classified on a scale from safe to critically endangered. One-fifth of all the species were found to be threatened, with 52 on average each year descending one step on the scale toward extinction. Extinction rates remain 100 to 1,000 times higher than before the global spread of humanity. *Conservation efforts made prior to the 2010 study were estimated to have slowed deterioration by at least one-fifth of what it otherwise would have been.* This is real progress, but still dismally short of stabilizing Earth's living environment. What would we think if told that the best efforts of (underfunded) medicine during a fatal pandemic had allowed only about 80 percent of the patients to die?

The remainder of the century will be a bottleneck of growing human impact on the environment and diminishment of biodiversity. We bear all of the responsibility of bringing ourselves and as much as possible of the rest of life through the bottleneck into a sustainable edenic existence. Our choice will be a profoundly moral one. Its fulfillment depends on knowledge still lacking and a sense of common decency still not felt. We alone among all species have grasped the reality of the living world, seen the beauty of nature, and given value to the individual. We alone have measured the quality of mercy among our own kind. Might we now extend the same concern to the living world that gave us birth?

IV

IDOLS OF THE MIND

HUMANITY'S INTELLECTUAL FRAILTIES

IDENTIFIED BY FRANCIS BACON, IN ONE OF

THE PRINCIPAL ACHIEVEMENTS OF THE FIRST

ENLIGHTENMENT, CAN NOW BE REDEFINED BY

SCIENTIFIC EXPLANATION.

12

Instinct

The French writer Jean Bruller (pen name Vercors) was on the right track when, in his 1952 novel *You Shall Know Them*, he declared, "All of man's troubles have arisen from the fact that we do not know what we are and do not agree on what we want to be."

In this part of our journey, I propose to come full circle and with the aid of general biology attempt to explain why human existence is such a mystery, and then expand on ways that mystery might be solved.

The human mind did not evolve as an externally guided progression toward either pure reason or emotional fulfillment. It remains as it has always been, an instrument of survival that employs both reason and emotion. It emerged in its present form from a labyrinth of large and small steps, in a series that is one out of millions possible. Each step in the labyrinth was an accident of mutation and natural selection acting on the alternative forms of genes that prescribe form and function of the brain and sensory system. By accident in the to and fro, it brought the genome to its present level. At each step the evolving genome could easily have veered onto one different pathway or another, hence specialization of the organism to a different kind of brain and sensory system. The chance of eventually reaching the human level would at each step have sharply declined.

The particular conglomerate of reason and emotion we call human nature was just one of many conceivable outcomes, an autonomously generated product reached first out of many kinds that could have achieved a brain and sensory system at the human level of capacity.

Such is the reason our self-image as a species has always been distorted by deep biases and misconceptions, the "idols" of superstition and imposture described by Francis Bacon four centuries ago. They were imposed upon us not by cultural accidents, the great philosopher said, but by the "general nature of the mind."

And so it has ever been. Confusion has always abounded. For example, as late as the 1970s the orientation of the social scientists was primarily toward the humanities. Their prevailing view was that human behavior is primarily or even entirely cultural, not biological in origin. There exists, the extremists among them claimed, no such thing as instinct and human

nature. By the end of the twentieth century the orientation flipped toward biology. Today, it is widely believed
that human behavior has a strong genetic component.
Instinct and human nature are real, although how deep
and forceful remains under discussion.

Both views, it turns out, are half wrong and half correct, at least in extremity. The paradox created, often
described as the nature-versus-nurture controversy, can
be solved by applying the modern concept of human
instinct, as follows.

Instinct in humans is basically the same as instinct in
animals. However, it is not the genetically fixed, invari-
ant behavior displayed by most animal species. A classi-
cal textbook example of the latter is territorial defense
by males of the three-spined stickleback, a fish found in
fresh and marine waters throughout the Northern Hemi-
sphere. During the breeding season each male stakes
out a small area which he defends from other males. At
the same time the male develops a bright red underside.
He attacks any other fish with a red belly, hence a rival
stickleback male, that enters his territory. Actually, the
response is even simpler than "other fish" implies. The
male has no need to recognize the full image of a real
fish in order to be activated. His relatively small brain
is programmed to respond just to the red belly. When
experimenters cut out pieces of wood in a near circle and

other unfishlike forms and painted a red spot on them, the models were attacked with equal vigor.

I once kept anole lizards from different West Indian islands in my laboratory, in order to study their territorial displays. The thumb-sized reptiles are everywhere abundant on trees and shrubs, where they prey on insects, spiders, and other small invertebrate animals. An adult male threatens rivals by lowering a flap of skin beneath his neck called the dewlap. The dewlap of each species has a different color, usually a shade of red or yellow or white; and males of the same species respond solely to that color. I found I needed only one male, not two, to get a territorial dewlap display. All I had to do was hold a mirror against the side of the terrarium. The resident male then displayed to his own image (resulting in a draw every time).

Infant sea turtles hatch from eggs buried in the sand of a beach by their mothers, who emerge from the sea solely for this purpose. Each of the hatchlings digs itself out and immediately crawls down to the sea, where it will spend the rest of its life. What attracts the newborn little animal is not, however, the many distinctive sights and odors emanating from the water's edge. The lure is instead the brighter light reflected off the surface of the water. When experimenters turned on an even brighter light close by, the baby turtle followed it, even when the light led it directly away from the sea.

Humans and other large-brained mammals are also guided by inherited key stimuli and instincts, but they are not nearly so rigid or simpleminded as those of lower animals. Instead, people in particular are ruled by what psychologists call prepared learning. What is inherited is the likelihood of learning one or a few alternative behaviors out of many possible. The strongest among the biased behaviors are shared across all cultures, even when they seem irrational and there are plenty of opportunities to make other choices.

I am a mild arachnophobe. I've tried on occasion but cannot bring myself to touch a large spider hanging in its web, even though I know it won't bite me, and even if it did, the bite would not be venomous. I've harbored this groundless fear ever since, at the age of eight, I was frightened by the sudden jerking behavior of a big garden spider of the orb-weaving genus *Araneus*. I had approached to examine the monster (so it seemed to me) closely, as it hung in sinister quietude in the center of its web, and was startled by its sudden response. Today I know its scientific name and a lot about its biology—as I should, having served for years as a curator of entomology at Harvard University's Museum of Comparative

Zoology. But I still won't—can't—touch large spiders hanging in their webs.

This kind of revulsion sometimes deepens in people to a full-blown phobia, characterized by panic, nausea, and an inability even to think rationally about the object of fear. Since I've just confessed a moderate, unjustified aversion, I'll also admit to the only true phobia I possess. I cannot bear nor will I tolerate under any imaginable condition having my arms pinned forcibly and my face covered. I remember with certainty the moment this response began. When eight years old, the year of the spider, I had a frightening eye operation. I was anesthetized in the nineteenth century manner—laid supine on an operating table without explanation or any I can recall, my arms then held down and a piece of cloth placed over my face into which ether was dripped. I was screaming as I struggled. Something deep within me must have said, Never again! To this day I have a "testing" fantasy of the phobia. I am held at gunpoint by an imaginary robber who tells me that he intends to tie my arms and place a hood over my face. My response in this scenario, and I believe it would be in reality as well, is to say, "No, you're not. So go ahead, shoot me." I would rather die than be tied up and hooded.

Phobias take a long time and a lot of therapy to remove. Yet they can be acquired with only a single experience, as I and so many others have personally discovered. The sudden appearance of something writhing on the ground, to take a second example, is enough for some people to acquire a phobia against snakes.

How could such overkill in learning confer any advantage? The clue is in the objects of the phobias themselves, which comprise mostly spiders, snakes, wolves, running water, closed spaces, and crowds of strangers. These were among the ancient perils of the prehumans and early human hunter-gatherers across millions of years. Our distant ancestors regularly faced injury or death while hunting for food too close to the edge of a ravine, or when they stepped carelessly on a venomous snake, or stumbled upon a raiding party of an enemy tribe. It was safest to learn fast, remember the event long and vividly, and act decisively without involving rational thought.

In contrast, automobiles, knives, guns, and the excessive consumption of dietary salt and sugar are among the leading causes of present-day mortality. Yet no inborn propensities to avoid them have evolved. The likely reason is the lack of time for evolution to have hardwired them into our brains.

Phobias are an extreme, but all behaviors acquired by prepared learning, having provided adaptive value during the ancestry of the human species, are part of human instinct. Yet most are also transmitted by culture from one generation to the next. All human social behavior is based on prepared learning, but the intensity of the bias varies from one case to another as a product of evolution by natural selection. For example, human beings are born gossips. We love the life stories of other people, and cannot be sated with too much such detail. Gossip is the means by which we learn and shape our social network. We devour novels and drama. But we have little or no interest in the life stories of animals— unless they are linked in some way to human stories. Dogs love others and yearn to return home, owls ponder, snakes sneak, and eagles thrill at the freedom of the open sky.

Human beings were made for music. Its thrill and rapture are picked up almost immediately by little children. However, the thrill (and scarcely ever rapture) of analytic mathematics comes a great deal more slowly and much later, if ever. Music served early humanity as a means of integrating societies and heightening the emotions of the people, but analytic mathematics never did. Early humans had the mental capacity to elaborate analytic mathematics, but not to love it. Only evolution by natural selection can create the need for bedrock instinctual love.

The driving force of natural selection has directed the convergence of cultural evolution among societies around the world. A classic synthesis of cultures made from the Human Relation Area Files in 1945 listed sixty-seven universals, including the following (selected here at random): athletic sports, bodily adornment, decorative art, etiquette, family feasting, folklore, funeral rites, hairstyles, incest taboos, inheritance rules, joking, and the propitiation to supernatural beings.

What we call human nature is the whole of our emotions and the preparedness in learning over which those emotions preside. Some writers have tried to deconstruct human nature into nonexistence. But it is real, tangible, and a process that exists in the structures of the brain. Decades of research have discovered that human nature is not the genes that prescribe the emotions and learning preparedness. It is not the cultural universals, which are its ultimate product. Human nature is the ensemble of hereditary regularities in mental development that bias cultural evolution in one direction as opposed to others and thus connect genes to culture in the brain of every person.

Among the more consequential hereditary biases in learning is the selection of the habitat in which people prefer to live. Adults are drawn to the kinds of environments in which they grew up and were shaped by their most formative experiences. For them, mountains, seashores, plains, and even deserts variously provide the habitats that give the greatest sense of familiarity and comfort. Having been raised myself mostly close to the Gulf of Mexico, I like best a flat, low plain that slopes down to the sea.

However, on a smaller scale within these panoramas, and for children not yet fully acculturated, laboratory experiments have yielded a different story. Volunteers from several countries with very different cultures were asked to evaluate photographs of a wide range of habitats where in fantasy they might live. The choices varied from dense forests to deserts, and other ecosystems in between. The preferred choice had three factors: the ideal vantage point is on a rise looking down, a vista of parkland comprising grassland sprinkled with trees and copses, and proximity to a body of water, whether stream, pond, lake, or ocean.

This archetype happens to be close to the actual savannas of Africa where our prehuman and early ancestors evolved over millions of years. Is it possible that the preference for the environment of the species remains as a residue of prepared learning? The "African savanna hypothesis," as it is called, is not at all a conjecture out of the blue. All mobile animal species, from the tiniest insects to elephants and lions, instinctively choose the habitats to which all the rest of their biology is best adapted. If they did not, they would be less likely to find a mate, the food on which they are dependent, or the means to avoid unfamiliar parasites and predators.

At the present time rural human populations around the world are imploding into cities. With any luck, their lives are improved by the better access to markets, schools, and medical centers. They also have a greater opportunity to support themselves and their families. But given a free choice, all else being equal, do they really prefer cities and suburbs as habitats? Because of the intense dynamism of urban ecology and the artifactual environment forced on them, it is impossible to say. So, to learn what people actually prefer and acquire when given a completely free choice, it is better to turn to those with a great deal of money. As landscape architects and high-end real estate agents will tell you, the rich prefer habitations set on a rise that looks out over parkland next to a body of water. None of these qualities have practical value, but people with sufficient means will pay any price to have them.

A few years ago I had dinner at the home of a distinguished and wealthy friend, who happened to be a firm believer in the brain as a blank slate, unencumbered by instinct. His home was a penthouse overlooking New York's Central Park. As we walked out onto the terrace, I noticed that its outer edge was lined with small potted trees. We looked down from there onto the distant grassy center of the park and one of its two artificial lakes. We agreed that the vista was all quite beautiful. Being a guest, I refrained from asking him the burning question: Why is it beautiful?

13

Religion

Rapture, a "joy excessive and sweet," as Spain's great mystic Saint Teresa of Ávila described it in her 1563–65 diary, can be achieved variously by music, religion—and hallucinogenic drugs such as the Amazonian religion-enhancer ayahuasca. Neurobiologists have tracked at least some of the peak experience of music to at least one cause, the release of the transmitter molecule dopamine within the striatum of the brain. The same biochemical reward system also mediates pleasure in food and sex. Because music began in Paleolithic times—bird-bone and ivory flutes date back more than thirty thousand years—and because it remains universal in hunter-gatherer societies around the world, it is reasonable to conclude that our loving devotion to it has been hard-wired by evolution in the human brain.

In almost all living societies, from hunter-gatherer to civilized-urban, there exists an intimate relation between music and religion. Are there genes for religiosity that prescribe a neural and biochemical mediation similar to that of music? Yes, says evidence from the relatively young discipline of the neuroscience of religion. The methods of inquiry include twin studies that measure the role of genetic variation, along with studies of hallucinogenic drugs that mimic religious experience. Also used are data concerning the impact on religiosity of brain lesions and other disorders, and, not least, the direct trajectory of the neural events tracked by brain imaging. Altogether, the results of the neuroscience of religion thus far suggest strongly that a religious instinct does indeed exist.

Of course there is far more to religion than its biological roots. Its history is as old or nearly so as that of humanity itself. The attempted resolution of its mysteries lies at the heart of philosophy. The purest, most general form of religion is expressed by theology, of which the central questions are the existence of God and God's personal relation to humanity. Deeply religious people want to find a way to approach and touch this deity—if not His literal transubstantiated flesh and blood in the Catholic manner, at least to ask Him for personal guidance and beneficence. Most also hope for life after death, passing into an astral world where they will join in bliss those who have gone before. Theological spirituality, in

short, seeks the bridge between the real and the supernatural. It dreams of God's dominion, where souls of the
Earthly dead live on together in peaceful eternity.

The brain was made for religion and religion for the
human brain. In every second of the believer's conscious
life religious belief plays multiple, mostly nurturing roles.
All the followers are unified into a vastly extended family, a metaphorical band of brothers and sisters, reliable,
obedient to one supreme law, and guaranteed immortality as the benefit of membership.

The deity is higher than any prophet, high priest,
imam, mystic saint, cult leader, president, emperor, dic-
tator, the lot. He is the final and forever alpha male, or
She the alpha female. Being supernatural and infinitely
powerful, the deity can perform miracles beyond the
reach of human understanding. Throughout prehistory
and most of history, people needed religion to explain the
occurrence of most phenomena around them. Torrential
rain and flooding, a lightning bolt streaking across the
sky, the sudden death of a child. God caused it. He or
She was the cause in the cause-and-effect required for
sanity. And the ways of God, albeit charged with meaning for our lives, are a mystery. With the coming of science, more and more natural phenomena have come to
be understood as effects linked to other analyzable phenomena, and supernatural explanations of cause-and-

effect have receded. But the deep, instinctive appeal of religion and religionlike ideology has remained.

The great religions are inspired by belief in an incorruptible deity—or multiple kinds of deities, who may also constitute an all-powerful family. They perform services invaluable to civilization. Their priests bring solemnity to the rites of passage through the cycle of life and death. They sacralize the basic tenets of civil and moral law, comfort the afflicted, and take care of the desperately poor. Inspired by their example, followers strive to be righteous in the sight of man and God. The churches over which they preside are centers of community life. When all else fails, these sacred places, where God dwells immanent on Earth, become ultimate refuges against the iniquities and tragedies of secular life. They and their ministers make more bearable tyranny, war, starvation, and the worst of natural catastrophes.

The great religions are also, and tragically, sources of ceaseless and unnecessary suffering. They are impediments to the grasp of reality needed to solve most social problems in the real world. Their exquisitely human flaw is tribalism. The instinctual force of tribalism in the genesis of religiosity is far stronger than the yearning for spirituality. People deeply need membership in a group, whether religious or secular. From a lifetime of emotional experience, they know that happiness, and indeed survival itself, require that they bond with others who share some amount of genetic kinship, language, moral beliefs, geographical location, social purpose, and dress code—preferably all of these but at least two or three for most purposes. It is tribalism, not the moral tenets and humanitarian thought of pure religion, that makes good people do bad things.

Unfortunately a religious group defines itself foremost by its creation story, the supernatural narrative that explains how humans came into existence. And this story is also the heart of tribalism. No matter how gentle and high-minded, or subtly explained, the core belief assures its members that God favors them above all others. It teaches that members of other religions worship the wrong gods, use wrong rituals, follow false prophets, and believe fantastic creation stories. There is no way around the soul-satisfying but cruel discrimination that organized religions by definition must practice among themselves. I doubt there ever has been an imam who suggested that his followers try Roman Catholicism or a priest who urged the reverse.

Acceptance of a particular creation story, and of accounts of miracles vouchsafed by it, is called the faith of the believer. Faith is biologically understandable as a Darwinian device for survival and increased reproduction. It is forged by the success of the tribe, the tribe is united by it when competing with other tribes, and it can be a key to success within the tribe for those members most effective in manipulating the faith to gain internal support. The unending conflicts that generated this powerful social practice were widespread through the Paleolithic Era and have continued unabated to the present time. In more secular societies faith tends to be transmuted into religionlike political ideologies. Sometimes the two great belief categories are combined. Hence, "God favors my political principles over yours, and my principles, not yours, favor God."

Religious faith offers enormous psychological benefit to the believers. It gives them an explanation for their existence. It makes them feel loved and protected above the members of every other tribal group. The price imposed by the gods and their priests in more primitive societies is unquestioning belief and submission. Throughout evolutionary time this bargain for the human soul was the only bond with the strength to hold the tribe together in both peace and war. It invested its members with a proud identity, legitimized rules of conduct, and explained the mysterious cycle of life and death.

For ages no tribe could survive unless the meaning of its existence was defined by a creation story. The price of the loss of faith was a hemorrhage of commitment, a weakening and dissipation of common purpose. In the early history of each tribe—late Iron Age for Judaeo-Christianity, and seventh century CE for Islam—the myth had to be set in stone in order to work. Once set, no part of it could be discarded. No doubts must be heard by the tribe. The only solution to an outmoded dogma was to finesse or conveniently forget it. Or, in the extreme, break away with a new, competing dogma.

Obviously no two creation stories can both be true. All of those invented by the many known thousands of religions and sects in fact have certainly been false. A great many educated citizens have realized that their own faiths are indeed false, or at least questionable in details. But they understand the rule attributed to the Roman stoic philosopher Seneca the Younger that religion is regarded by the common people as true, by the wise as false, and by rulers as useful.

Scientists by nature tend to be cautious in anything they say about religion, even when expressing skepticism. The distinguished physiologist Anton (Ajax) J. Carlson, when asked what he thought of the 1950 ex cathedra (that is, infallible) pronouncement by Pius XII that the Virgin Mary ascended bodily into heaven, is reported to have responded that he couldn't be sure because he wasn't there, but of one thing he was certain, that she passed out at thirty thousand feet.

Might it be better just to leave this vexatious matter alone? Not deny, just forget? After all, the great majority of people in the world are sort of getting along, more or less. However, negligence in the matter is dangerous, both short-term and long-term. National wars may have subsided, obviously due to the fear of their possibly catastrophic outcomes to both sides. But insurgencies, civil wars, and terrorism have not. The principal driving force of mass murders committed during them is tribalism, and the central rationale for lethal tribalism is sectarian religion—in particular the conflict between those faithful to different myths. At the time of writing the civilized world flinches before the brutal struggles between Shiites and Sunnis, the murder of Ahmadiyya Muslims in Pakistan's cities by other Muslims, and the slaughter of Muslims by Buddhist-led "extremists" in Myanmar. Even the blocking by ultra-Orthodox Jews of liberal Jewish women from the Western Wall is a menacing early symptom of the same social pathology.

Religious warriors are not an anomaly. It is a mistake to classify believers of particular religious and dogmatic religionlike ideologies into two groups, moderate versus extremist. The true cause of hatred and violence is faith versus faith, an outward expression of the ancient instinct of tribalism. Faith is the one thing that makes otherwise good people do bad things. Nowhere do people tolerate attacks on their person, their family, their country—or their creation myth. In America, for example, it is possible in most places to openly debate different views on religious spirituality—including the nature and even the existence of God, providing it is in the context of theology and philosophy. But it is forbidden to question closely, if at all, the creation myth—the faith—of another person or group, no matter how absurd. To disparage anything in someone else's sacred creation myth is "religious bigotry." It is taken as the equivalent of a personal threat.

Another way of expressing the history of religion is that faith has hijacked religious spirituality. The prophets and leaders of organized religions, consciously or not, have put spirituality in the service of groups defined by their creation myths. Awe-inspiring ceremonies and sacred rites and rituals and sacrifices are given the deity in return for worldly security and the promise of immortality. As part of the exchange the deity must also make correct moral decisions. Within the Christian faith, among most of the denominational tribes, God is obliged to be against one or more of the following: homosexuality, artificial contraception, female bishops, and evolution.

The Founding Fathers of the United States understood the risk of tribal religious conflict very well.

George Washington observed, "Of all the animosities which have existed among mankind those which are caused by difference of sentiments in religion appear to be the most inveterate and distressing and ought most to be deprecated." James Madison agreed, noting the "torrents of blood" that result from religious competition. John Adams insisted that "the government of the United States is not in any sense founded on the Christian religion." America has slipped a bit since then. It has become almost mandatory for political leaders to assure the electorate that they have a faith, even, as for the Mormonism of Mitt Romney, if it looks ridiculous to the great majority. Presidents often listen to the counsel of Christian advisers. The phrase "under God" was introduced into the Pledge of Allegiance in 1954, and today no major political candidate would dare suggest it be removed.

Most serious writers on religion conflate the transcendent quest for meaning with the tribalistic defense of creation myths. They accept, or fear to deny, the existence of a personal deity. They read into the creation myths humanity's effort to communicate with the deity, as part of the search for an uncorrupted life now and beyond death. Intellectual compromisers one and all, they include liberal theologians of the Niebuhr school, philosophers battening on learned ambiguity, literary admirers of C. S. Lewis, and others persuaded, after deep thought, that there must be Something Out There. They tend to be unconscious of prehistory and the biological evolution of human instinct, both of which beg to shed light on this very important subject.

The compromisers face an insoluble problem, which the great, conflicted nineteenth century Danish philosopher Søren Kierkegaard called the Absolute Paradox. Dogmas forced on believers, he said, are not just impossible but incomprehensible—hence absurd. What Kierkegaard had in mind in particular was the core of the Christian creation myth. "The Absurd is that the eternal truth has come to exist, that God has come to exist, is born, has grown up and so on, and has become just like a person, impossible to tell apart from another person." It was incomprehensible, even if declared true, Kierkegaard continued, that God as Christ entered into the physical world in order to suffer, leaving martyrs to suffer for real.

The Absolute Paradox tears at all in every religion who seek an honest resolution of body and soul. It is the inability to conceive of an all-knowing divinity who created a hundred billion galaxies, yet whose humanlike emotions include feelings of pleasure, love, generosity, vindictiveness, and a consistent and puzzling lack of concern for the horrific things Earth-dwellers endure under the deity's rule. To explain that "God is testing our faith" and "God moves in mysterious ways" doesn't cut it.

As Carl Jung once said, some problems can never be solved, only outgrown. And so it must be for the Absolute Paradox. There is no solution because there is nothing to solve. The problem is not in the nature or even in the existence of God. It is in the biological origins of human existence and in the nature of the human mind, and what made us the evolutionary pinnacle of the biosphere. The best way to live in this real world is to free ourselves of demons and tribal gods.

14

Free Will

Neuroscientists who work on the human brain seldom mention free will. Most consider it a subject better left, at least for the time being, to philosophers. "We will attend to it when we're ready and have time," they seem to say. Meanwhile, their sights are set on the brighter and more realistically conceived grail of science, the physical basis of consciousness, of which free will is a part. No scientific quest is more important to humanity than to nail the phantom of conscious thought. Everyone, scientists, philosophers, and religious believers alike, can agree with the neurobiologist Gerald Edelman that "consciousness is the guarantor of all we hold to be human and precious. Its permanent loss is considered to be the equivalent of death, even if the body persists in its vital signs."

The physical basis of consciousness won't be an easy phenomenon to grasp. The human brain is the most complex system known in the Universe, either organic or inorganic. Each of the billions of nerve cells (neurons) composing its functional part forms synapses and communicates with an average of ten thousand others. Each launches messages along its own axon pathway, using an individual digital code of membrane firing patterns. The brain is organized into regions, nuclei, and staging centers that divide functions among them. The parts respond in different ways to hormones and sensory stimuli originating from outside the brain, while sensory and motor neurons all over the body communicate so intimately with the brain as to be virtually a part of it.

Half of the twenty thousand to twenty-five thousand genes of the entire human genetic code participate in one manner or other in the prescription of the brain-mind system. This amount of commitment has resulted from one of the most rapid evolutionary changes known in any advanced organ system of the biosphere. It entailed a more than twofold increase in brain size across three million years, from at or below 600cc in the australopith prehuman ancestor to 680cc in *Homo habilis*, thence to about 1,400cc in modern *Homo sapiens*.

Philosophers have labored off and on for over two thousand years to explain consciousness. Of course they have, it's their job. Innocent of biology, however, they have for the most part understandably gotten nowhere. I don't believe it too harsh to say that the history of philosophy when boiled down consists mostly of failed models of the brain. A few of the modern neurophilosophers such as Patricia Churchland and Daniel Dennett have made a splendid effort to interpret the findings of neuroscience research as these become available. They have helped others to understand, for example, the ancillary nature of morality and rational thought. Others, especially those of poststructuralist bent, are more retrograde. They doubt that the "reductionist" or "objectivist" program of the brain researchers will ever succeed in explaining the core of consciousness. Even if it has a material basis, subjectivity in this view is beyond the reach of science. To make their argument, the mysterians (as they are sometimes called) point to the qualia, the subtle, almost inexpressible feelings we experience about sensory input. For example, "red" we know from physics, but what are the deeper sensations of "redness"? So what can the scientists ever hope to tell us in larger scale about free will, or about the soul, which for religious thinkers at least is the ultimate of ineffability?

The procedure of the more skeptical philosophers is top-down and introspective—thinking about how we think, then adducing explanations, or else discovering reasons why there can be no explanation. They describe the phenomena and they provide thought-provoking examples. They conclude that there is something fundamentally different from ordinary reality in the conscious mind. Whatever that may be, it is better left to philosophers and poets.

Neuroscientists, who are relentlessly bottom-up as opposed to top-down, will have none of this. They have no illusions about the difficulty of the task, understanding that mountains are not provided with escalators built for dreamers. They agree with Darwin that the mind is a citadel that cannot be taken by frontal assault. They have set out instead to break through to its inner recesses with multiple probes along the ramparts, opening breaches here and there, and by technical ingenuity and force enter and explore wherever space is found to maneuver.

You have to have faith to be a neuroscientist. Who knows where consciousness and free will may be hidden—assuming they even exist as integral processes and entities? Will they emerge in time, metamorphosing from the data like a butterfly from a caterpillar, the image filling us like Keats's men around Balboa with a wild surmise? Meanwhile, neuroscience, primarily because of its relevance to medicine, has grown rich. Its research projects are growing on budgets of hundreds of millions to billions each year. (In the science trade it's called Big Science.) The same surge has occurred, successfully, in cancer research, the space shuttle, and experimental particle physics.

As I write, neuroscientists have begun the direct assault Darwin called impossible. It envisions an effort called the Brain Activity Map (BAM) Project, conceived by key government institutes in the United States, including the National Institutes of Health and National Science Foundation, in collaboration with the Allen Institute for Brain Science, and endorsed as government policy by President Obama. The program, if successfully funded, will parallel in magnitude the Human Genome Project, which was the biology moon shot completed in 2003. Its goal is nothing less than a map of the activity of every neuron in real time. Much of the technology will have to be developed on the job.

The basic goal of activity mapping is to connect all of the processes of thought—rational and emotional, conscious, preconscious, and unconscious, held both still and moving through time—to a physical base. It won't come easy. Bite into a lemon, fall into bed, recall a departed friend, watch the sun sink beyond the western sea. Each episode comprises mass neuronal activity so elaborate, so little of which as yet been seen, we cannot even conceive it, much less write it down as a repertory of firing cells.

Skepticism concerning BAM is rife among scientists, but that is nothing new. The same resistance was also the case for the Human Genome Project and much of space exploration being conducted by NASA. An added incentive to push ahead is the practical application mapping will have for medicine—in particular the cellular and molecular foundations of mental illness and the discovery of deleterious mutations even before the symptoms are expressed.

Assuming that BAM or other, similar enterprises are successful, how might they solve the riddle of consciousness and free will? I suggest that the solution will come relatively early in the functional mapping program rather than as a grand finale at the end, and providing neurobiology remains favored as Big Science. In evidence is the large amount of information already archived in brain studies, and especially as it is combined with the principles of evolutionary biology.

There are several reasons for optimism in the search for an early solution. First is the gradual emergence of consciousness during evolution. The extraordinarily high human level was not reached suddenly, like a light turned on with the flick of a switch. The gradual albeit rapid increase in brain size leading up from the habiline prehumans to *Homo sapiens* suggests that consciousness evolved in steps, in a manner similar to those of other complex biological systems—the eukaryotic cell, for example, or the animal eye, or colonial life in insects.

It should then be possible to track the steps leading to human consciousness through studies of animal species that have come partway to the human level. The mouse has been a prominent model in early brain-mapping research, and will continue to be productive. This species has considerable technical advantages, including convenient laboratory rearing (for a mammal) and a strong supporting foundation of prior genetic and neuroscience research. A closer approach to the actual sequence can be made, however, by also adding humanity's closest phylogenetic relatives among the Old World primates, from lemurs and galagos at the more primitive end to rhesus macaques and chimpanzees at the higher end. The comparison would reveal what neural circuits and activities were attained by nonhuman species, and when and in what sequence. The data obtained might detect, even at a relatively early stage of research, the neurobiological traits that are uniquely human.

The second point of entry into the realm of consciousness and free will is the identification of emergent phenomena—entities and processes that come into existence only with the joining of preexisting entities and processes. They will be found, if the results of current research are indicative, in the linkage and synchronized activity of various parts of both the sensory system and the brain.

Meanwhile, the nervous system can be usefully conceived as a superbly well organized superorganism built upon a division of labor and specialization in the society of cells—around which the body plays a primarily supportive role. An analog, if you will, is to be found in a queen ant or termite, and her supporting swarm of workers. Each worker on its own is relatively stupid. It follows a program of blind, untutored instinct, which is subject to only a small amount of flexibility in its expression. The program directs the worker to specialize on one or two tasks at a time, and to change programs in a particular sequence—typically nurse to builder and guard to forager—as the worker ages. All the workers together are in contrast brilliant. They address all needed tasks simultaneously, and can shift the weight of their effort to meet potentially lethal emergencies such as flooding, starvation, and attacks by enemy colonies. This comparison should not be considered a stretch. Something like it has been a common theme in serious literature as far back as Douglas Hofstadter's 1979 classic *Gödel, Escher, Bach: An Eternal Golden Braid.*

A strong additional advantage is the narrowness of the range of human perception. Our sight, hearing, and other senses impart the feeling that we are aware of almost everything around us in both space and time. Yet, as I've emphasized earlier, we are aware of only minute slivers of space-time, and even less of the energy fields, in which we exist. The conscious mind is a map of our awareness in the intersection only of those parts of the continua we happen to occupy. It allows us to see and know those events that most affect our survival in the real world, or, more precisely, the real world in which our prehuman ancestry evolved. To understand sensory information and the passage of time is to understand a large part of consciousness itself. Advance in this direction might prove easier than previously assumed.

The final reason I'd like to suggest for optimism is the human necessity for confabulation. Our minds consist of storytelling. In each instant of present time, a flood of real-world information flows into our senses. Added to the severe limitation of the senses is the fact that the information they receive far exceeds what the brain can process. To augment this fraction, we summon the stories of past events for context and meaning. We compare them with the unfolding past to apply the decisions that were made back in time, variously right or wrong. Then we look forward to create—not just to recall this time—multiple competing scenarios. These are weighed against one another by the suppressing or intensifying effect imposed by aroused emotional centers. A choice is made in the unconscious centers of the brain, it turns out from recent studies, several seconds before the decision arrives in the conscious part.

Conscious mental life is built entirely from confabulation. It is a constant review of stories experienced in the past and competing stories invented for the future. By necessity most conform to the present real world as best it can be processed by our rather paltry senses. Memories of past episodes are repeated for pleasure, for rehearsal, for planning, or for various combinations of the three. Some of the memories are altered into abstractions and metaphors, the higher generic units that increase the speed and effectiveness of the conscious process.

Most conscious activity contains elements of social interactions. We are fascinated by the histories and emotional responses of others. We play games, both imaginary and real, based on the reading of intention and probable response. Sophisticated stories at this level require a big brain housing vast memory banks. In the human world that capacity evolved long ago as an aid to survival.

If consciousness has a material basis, can the same be true for free will? Put another way, what if anything in the manifold activities of the brain could possibly pull away from the brain's machinery in order to create scenarios and make decisions of its own? The answer is of course the self. And what would that be? And where is it? The self cannot exist as a paranormal being living on its own within the brain. It is instead the central dramatic character of the confabulated scenarios. In these stories it is always on center stage, if not as participant then as observer and commentator, because that is where all of the sensory information arrives and is integrated. The stories that compose the conscious mind cannot be taken away from the mind's physical neurobiological system, which serves as script writer, director, and cast combined. The self, despite the illusion of its independence created in the scenarios, is part of the anatomy and physiology of the body.

The power to explain consciousness will, however, always be limited. Suppose that the neuroscientists somehow successfully learned all of the processes of the brain of a single person in detail. Could they then explain the mind of that particular individual? No, not even close. It would require opening up the immense store of the brain's particular memories, both those images and events available to immediate recall and others buried deep in the unconscious. And if such a feat were possible, even in a limited way, its accomplishment would modify the memories and the emotional centers that respond to the memories, causing a new mind to emerge.

Then there is the element of chance. The body and brain comprise legions of communicating cells, which shift in discordant patterns that cannot even be imagined by the conscious minds they compose. The cells are bombarded every instant by outside stimuli unpredictable by human intelligence. Any one of these events can entrain a cascade of changes in local neural patterns, and scenarios of individual minds changed by them are all but infinite in detail. The content is dynamic, changing instant by instant in accordance with the unique history and physiology of the individual.

Because the individual mind cannot be fully described by itself or by any separate researcher, the self—celebrated star player in the scenarios of consciousness—can go on passionately believing in its independence and free will. And that is a very fortunate Darwinian circumstance. Confidence in free will is biologically adaptive. Without it the conscious mind, at best a fragile dark window on the real world, would be cursed by fatalism. Like a prisoner confined for life to solitary confinement, deprived of any freedom to explore and starving for surprise, it would deteriorate.

So, does free will exist? Yes, if not in ultimate reality, then at least in the operational sense necessary for sanity and thereby for the perpetuation of the human species.

V

A HUMAN FUTURE

IN THE TECHNOSCIENTIFIC AGE, FREEDOM HAS
ACQUIRED A NEW MEANING. LIKE AN ADULT
EMERGING FROM CHILDHOOD, WE HAVE A
VASTLY WIDER RANGE OF CHOICES BUT ALSO A
COMPARABLY LARGER NUMBER OF RISKS AND
RESPONSIBILITIES.

15

Alone and Free in the Universe

What does the story of our species tell us? By this I mean the narrative made visible by science, not the archaic version soaked in religion and ideology. I believe the evidence is massive enough and clear enough to tell us this much: We were created not by a supernatural intelligence but by chance and necessity as one species out of millions of species in Earth's biosphere. Hope and wish for otherwise as we will, there is no evidence of an external grace shining down upon us, no demonstrable destiny or purpose assigned us, no second life vouchsafed us for the end of the present one. We are, it seems, completely alone. And that in my opinion is a very good thing. It means we are completely free. As a result we can more easily diagnose the etiology of the irrational beliefs that so unjustifiably divide us. Laid before us are new options scarcely dreamed of in earlier ages. They empower us to address with more confidence the greatest goal of all time, the unity of the human race.

The prerequisite for attaining the goal is an accurate self-understanding. So, what is the meaning of human existence? I've suggested that it is the epic of the species, begun in biological evolution and prehistory, passed into recorded history, and urgently now, day by day, faster and faster into the indefinite future, it is also what we will choose to become.

To speak of human existence is to bring into better focus the difference between the humanities and science. The humanities address in fine detail all the ways human beings relate to one another and to the environment, the latter including plants and animals of aesthetic and practical importance. Science addresses everything else. The self-contained worldview of the humanities describes the *human condition*—but not why it is the one thing and not another. The scientific worldview is vastly larger. It encompasses the meaning of *human existence*— the general principles of the human condition, where the species fits in the Universe, and why it exists in the first place.

Humanity arose as an accident of evolution, a product of random mutation and natural selection. Our species was just one end point out of many twists and turns in a single lineage of Old World primates (prosimians, monkeys, apes, humans) of which there are today several hundred other native species, each a product of its own twists and turns. We might easily have remained just another australopith with an ape-sized brain, collecting fruit and grabbing at fish, eventually to suffer extinction like other australopiths.

For the four hundred million years that large animals have occupied the land, *Homo sapiens* has been the only one to evolve intelligence high enough to create a civilization. Our genetically nearest relatives, the chimpanzees, today represented by two species (the common chimpanzee and the bonobo), came closest. The human and chimpanzee lineages split from a common stock in Africa about six million years ago. Roughly two hundred thousand generations have passed, plenty of time for natural selection to force a series of major genetic changes. The prehumans possessed certain advantages that biased the direction of their subsequent evolution. These included, at the beginning, a partly arboreal life and free use of the forelimbs that went with it. This archaic condition was then altered to a primarily ground-dwelling life. Also in place as biasing conditions were large-brained ancestors and an immense continent with a mostly equitable climate and extensive grassland interspersed with open dry forest. In later years the favoring preconditions included frequent ground fires that promoted fresh growth in

herbaceous and shrubby plants. Also and more importantly, the fires made possible an eventual dietary shift
to cooked meat. This rare combination of circumstances
during the evolutionary run-up, combined with luck (no
devastating climate change, volcanic eruptions, or severe
pandemics), rolled the dice in favor of the early humans.

Godlike, their descendants have saturated a large part
of Earth, and altered to varying degree the remainder.
We have become the mind of the planet and perhaps our
entire corner of the galaxy as well. We can do with Earth
what we please. We chatter constantly about destroying
it—by nuclear war, climate change, an apocalyptic Second Coming foretold by Holy Scripture.

Human beings are not wicked by nature. We have
enough intelligence, goodwill, generosity, and enterprise
to turn Earth into a paradise both for ourselves and for
the biosphere that gave us birth. We can plausibly accomplish that goal, at least be well on the way, by the end of
the present century. The problem holding everything up
thus far is that Homo sapiens is an innately dysfunctional
species. We are hampered by the Paleolithic Curse:
genetic adaptations that worked very well for millions of
years of hunter-gatherer existence but are increasingly a
hindrance in a globally urban and technoscientific society. We seem unable to stabilize either economic policies or the means of governance higher than the level

of a village. Further, the great majority of people world-wide remain in the thrall of tribal organized religions, led by men who claim supernatural power in order to compete for the obedience and resources of the faithful. We are addicted to tribal conflict, which is harmless and entertaining if sublimated into team sports, but deadly when expressed as real-world ethnic, religious, and ideological struggles. There are other hereditary biases. Too paralyzed with self-absorption to protect the rest of life, we continue to tear down the natural environment, our species' irreplaceable and most precious heritage. And it is still taboo to bring up population policies aiming for an optimum people density, geographic distribution, and age distribution. The idea sounds "fascist," and in any case can be deferred for another generation or two—we hope.

Our species' dysfunction has produced the hereditary myopia of which we are all uncomfortably familiar. People find it hard to care about other people beyond their own tribe or country, and even then past one or two generations. It is harder still to be concerned about animal species—except for dogs, horses, and others of the very few we have domesticated to be our servile companions.

Our leaders, religious, political, and business, mostly accept supernatural explanations of the human existence. Even if privately skeptical, they have little interest in opposing religious leaders and unnecessarily stirring up the populace, from whom they draw power and privilege. Scientists who might contribute to a more realistic worldview are especially disappointing. Largely yeomen, they are intellectual dwarves content to stay within the narrow specialties for which they were trained and are paid.

Some of the dysfunction of course comes from the youthful state of global civilization, which is still a work in progress. But the greater part is due simply to the fact that our brains are poorly wired. Hereditary human nature is the genetic legacy of our prehuman and Paleolithic past—the "indelible stamp of our lowly origin" as identified by Charles Darwin, first in anatomy (*The Descent of Man*, 1871) and then in the facial signals of emotion (*The Expression of the Emotions in Man and Animals*, 1872). Evolutionary psychologists have pressed on to explain the role of biological evolution in gender differences, child mental development, status ranking, tribal aggression, and even dietary choice.

As I've suggested in previous writing, the chain of causation runs yet deeper, extending all the way to the level of the biological organization on which natural selection works. Selfish activity within the group provides competitive advantage but is commonly destructive to the group as a whole. Working in the opposite direction from individual-level selection is group selection—group versus group. When an individual is cooperative and altruistic, this reduces his advantage in competition to a comparable degree with other members but increases the survival and reproduction rate of the group as a whole. In a nutshell, individual selection favors what we call sin and group selection favors virtue. The result is the internal conflict of conscience that afflicts all but psychopaths, estimated fortunately to make up only 1 to 4 percent of the population.

The products of the opposing two vectors in natural selection are hardwired in our emotions and reasoning, and cannot be erased. Internal conflict is not a personal irregularity but a timeless human quality. No such conflict exists or can exist in an eagle, fox, or spider, for example, whose traits were born solely of individual selection, or a worker ant, whose social traits were shaped entirely by group selection.

The internal conflict in conscience caused by competing levels of natural selection is more than just an arcane subject for theoretical biologists to ponder. It is not the presence of good and evil tearing at one another in our breasts. It is a biological trait fundamental to understanding the human condition, and necessary for survival of the species. The opposed selection pressures during the genetic evolution of prehumans produced an unstable mix of innate emotional response. They created a mind that is continuously and kaleidoscopically shifting in mood—variously proud, aggressive, competitive, angry, vengeful, venal, treacherous, curious, adventurous, tribal, brave, humble, patriotic, empathetic, and loving. All normal humans are both ignoble and noble, often in close alternation, sometimes simultaneously.

The instability of the emotions is a quality we should wish to keep. It is the essence of the human character, and the source of our creativity. We need to understand ourselves in both evolutionary and psychological terms in order to plan a more rational, catastrophe-proof future. We must learn to behave, but let us never even think of domesticating human nature.

Biologists have created the very useful concept of the tolerable parasite load, defined as onerous but not unbearable. Almost all species of plants and animals carry parasites, which by definition are other species that live on or inside their bodies and in most instances take some little part of the hosts without killing them. Parasites, in a phrase, are predators that eat prey in units of less than one. Tolerable parasites are those that have evolved to ensure their own survival and reproduction but at the same time with minimum pain and cost to the host. It would be a mistake for an individual to try to eliminate all of its tolerable parasites. The cost would be too great in time and too disruptive to its own bodily functions. If you doubt this principle, think about what it would take to exterminate the almost-microscopic demodex mites that may (roughly 50 percent probability) live at this moment at the base of your eyebrow hairs. Also, consider the millions of unfriendly bacteria dwelling alongside the friendly ones in the nutritionally rich liquids of your mouth.

Destructive inborn traits of social life can be viewed as a parallel of the physical presence of parasitic organisms, and the cultural diminishment of their impact as the lessening of a tolerable dogma load. One obvious example of the latter is blind faith in supernatural creation stories. Of course in most parts of the world today, moderating the dogma load would be difficult, even dangerous. The stories are harnessed to both tribal rule by means of subordination of the faithful and their assumption of religious superiority over believers of rival creation stories. To examine each of the stories in detail objectively and to spell out their known historical origins would be a good start, and one that has begun (albeit slowly and carefully) in many scholarly disciplines. A second step, granted an unrealistic one, would be to ask the leaders of each religion and sect, assisted by theologians, to publicly defend the supernatural details of their faiths in competition with other faiths and aided by natural-cause and historical analysis.

It has been the universal practice to denounce such challenges to the core doctrines of particular faiths as blasphemous. Yet it would be far from irrational in today's better-informed world to reverse the practice, and charge with blasphemy any religious or political leader who claims to speak with or on behalf of God. The idea is to place the personal dignity of the believer above the dignity of the belief that demands his unquestioning obedience. It might eventually be possible to hold seminars on the historical Jesus in evangelical churches, and even to publish images of Muhammad without risking death.

That would be a true cry of freedom. The same practice might be adopted for dogmatic political ideologies, of which we have altogether too many around the world. The reasoning behind these secular religions is always the same, a proposition considered to be logically true followed by top-down explanation and a handpicked checklist of evidence asserted to be supportive. Zealots and dictators alike would feel their strength draining away if they were asked to explain their assumptions ("speak clearly, please") and verify their core beliefs.

Among the most virulent of all such cultural parasite-equivalents is the religion-based denial of organic evolution. About one-half of Americans (46 percent in 2013, up from 44 percent in 1980), most of whom are evangelical Christians, together with a comparable fraction of Muslims worldwide, believe that no such process has ever occurred. As Creationists, they insist that God created humankind and the rest of life in one to several magical mega-strokes. Their minds are closed to the overwhelming mass of factual demonstrations of evolution, which is increasingly interlocked across every level of biological organization from molecules to ecosystem and the geography of biodiversity. They ignore, or more precisely they call it virtue to remain ignorant of, ongoing evolution observed in the field and even traced to the genes involved. Also looked past are new species created in the laboratory. To Creationists, evolution is at best just an unproven theory. To a few, it is an idea invented by Satan and transmitted through Darwin and later scientists in order to mislead humanity. When I was a small boy attending an evangelical church in Florida, I was taught that the secular agents of Satan are extremely bright and determined, but liars all, man and woman, and so no matter what I heard I must stick my fingers in my ears and hold fast to the true faith.

We are all free in a democracy to believe whatever we wish, so why call any opinion such as Creationism a virulent cultural parasite-equivalent? Because it represents a triumph of blind religious faith over carefully tested fact. It is not a conception of reality forged by evidence and logical judgment. Instead, it is part of the price of admission to a religious tribe. Faith is the evidence given of a person's submission to a particular god, and even then not to the deity directly but to other humans who claim to represent the god.

The cost to society as a whole of the bowed head has been enormous. Evolution is a fundamental process of the Universe, not just in living organisms but everywhere, at every level. Its analysis is vital to biology, including medicine, microbiology, and agronomy. Furthermore psychology, anthropology, and even the history of religion itself make no sense without evolution as the key component followed through the passage of time. The explicit denial of evolution presented as a part of a "creation science" is an outright falsehood, the adult equivalent of plugging one's ears, and a deficit to any society that chooses to acquiesce in this manner to a fundamentalist faith.

Granted there are positive consequences to blind faith. It binds groups more strongly, and provides comfort to their members. It promotes charity and law-abiding behavior. Possibly the dogma load is made more tolerable by these services. Still, the ultimate force driving blind faith is not a divine afflatus. It is instead certification of membership in a group. The welfare of the group and defense of its territory is biological, not supernatural in origin. Except in theologically repressive societies, it has proved easy for individuals to shift religion, marry across religions, and even drop them entirely without loss of morality or, of equal importance, the capacity for wonder.

There are other archaic misconceptions outside religion that have weakened culture, albeit with a more logical and honorable rationale. The most important is the belief that the two great branches of learning—science and the humanities—are intellectually independent of each other. And more, the farther apart they are kept, the better.

I've argued here that while scientific knowledge and technology continue to grow exponentially, doubling every one to two decades according to discipline, the rate of increase will inevitably slow. Original discoveries, having generated vast knowledge, will ease off and begin to decline in number. Within decades, knowledge within the technoscientific culture will of course be enormous compared to that of the present, but also the same everywhere in the world. What will continue to evolve and diversify indefinitely are the humanities. If our species can be said to have a soul, it lives in the humanities.

Yet this great branch of learning, including the creative arts and their scholarly criticism, is still hampered by the severe and widely unappreciated limitations of the sensory world in which the human mind exists. We are primarily audiovisual and unaware of the world of taste and smell in which most of the millions of other species exist. We are entirely oblivious to the electrical and magnetic fields used by a few animals for orientation and communication. Even in our own world of sight and sound we are relatively close to blind and deaf, able to perceive directly no more than minute segments of the electromagnetic spectrum, nor the full range of compression frequencies that surge past us through earth, air, and water.

And that is just the start. Although the details of the creative arts are potentially infinite, the archetypes and instinct they are designed to exemplify are in reality very few. The ensemble of emotions that produce them, even the most powerful, are sparse—fewer in number than, say, the instruments of a full orchestra. Creative artists and humanities scholars by and large have little grasp of the otherwise immense continuum of space-time on Earth, in both its living and nonliving parts, and still less in the Solar System and the Universe beyond. They have the correct perception of *Homo sapiens* as a very distinctive species, but spend little time wondering what that means or why it is so.

Science and the humanities, it is true, are fundamentally different from each other in what they say and do. But they are complementary to each other in origin, and they arise from the same creative processes in the human brain. If the heuristic and analytic power of science can be joined with the introspective creativity of the humanities, human existence will rise to an infinitely more productive and interesting meaning.

Appendix

The Limitations of Inclusive Fitness

Because of the importance of genetic theory used to explain the biological origins of altruism and advanced social organization, and the much-publicized recent controversy surrounding it, I have included here a recent analysis of the theory of inclusive fitness, and the reason it should be replaced by data-based population genetics. The material presented is of a previously published research report, with the mathematical analyses and references deleted. The article was subject to intense expert review prior to publication.

Reference: "Limitations of Inclusive Fitness," by Benjamin Allen, Martin A. Nowak, and Edward O. Wilson, *Proceedings of the National Academy of Sciences USA*, volume 110, number 50, pages 20135–20139 (2013).

Significance

Inclusive fitness theory is the idea that the evolutionary success of a trait can be calculated as a sum of fitness effects multiplied by relatedness coefficients. Despite recent mathematical analyses demonstrating the limitations of this approach, its adherents claim that it is as general as the theory of natural selection itself. This claim is based on using linear regression to split an individual's fitness into components due to self and others. We show that this regression method is useless for the prediction or interpretation of evolutionary processes. In particular, it fails to distinguish between correlation and causation, leading to misinterpretations of simple scenarios. The weaknesses of the regression method underscore the limitations of inclusive fitness theory in general.

Until recently, inclusive fitness has been widely accepted as a general method to explain the evolution of social behavior. Affirming and expanding earlier criticism, we demonstrate that inclusive fitness is instead a limited concept, which exists only for a small subset of evolutionary processes. Inclusive fitness assumes that personal fitness is the sum of additive components caused by individual actions. This assumption does not hold for the majority of evolutionary processes or scenarios. To sidestep this limitation, inclusive fitness theorists have proposed a method using linear regression. On the basis of this method, it is claimed that inclusive fitness theory (i) predicts the direction of allele frequency changes, (ii) reveals the reasons for these changes, (iii) is as general as natural selection, and (iv) provides a universal design principle for evolution. In this paper we evaluate these claims, and show that all of them are unfounded. If the objective is to analyze whether mutations that modify social behavior are favored or opposed by natural selection, then no aspect of inclusive fitness theory is needed.

Inclusive fitness theory is an approach to accounting for fitness effects in social evolution. It was introduced in 1964 by W. D. Hamilton, who showed that, under certain circumstances, evolution selects for organisms with the highest inclusive fitness. This result has been interpreted as a design principle: evolved organisms act as if to maximize their inclusive fitness.

Hamilton defined inclusive fitness as follows:

Inclusive fitness may be imagined as the personal fitness which an individual actually expresses in its production of adult offspring as it becomes after it has been first stripped and then augmented in a certain way. It is stripped of all components which can be considered as due to the individual's social environment, leaving the fitness which he would express if not exposed to any of the harms or benefits of that environment. This quantity is then augmented by certain fractions of the quantities of harm and benefit which the individual himself causes to the fitnesses of his neighbours. The fractions in question are simply the coefficients of relationship appropriate to the neighbours whom he affects: unity for clonal individuals, one-half for sibs, one-quarter for half-sibs, one-eighth for cousins ... and finally zero for all neighbours whose relationship can be considered negligibly small.

Although modern formulations of inclusive fitness theory use different relatedness coefficients, all other aspects of Hamilton's definition remain intact.

The crucial point here is that it is assumed that personal fitness can be subdivided into additive components caused by individual actions. The personal fitness of a focal individual is stripped of all components that are due to the "social environment." This means we have to subtract from the personal fitness of an individual every effect due to other individuals. Subsequently we have to calculate how the focal individual affects the personal fitnesses of all other individuals in the population. In both cases we must assume that personal fitness can be expressed as a sum of components caused by individual actions. Inclusive fitness is the effect of the action on the actor plus the effects of the action on others multiplied in each case by the relatedness between the actor and the others.

It is immediately obvious that the additivity assumption, which is essential for the concept of inclusive fitness, need not hold in general. For example, the personal fitness of an individual can be a nonlinear function of the actions of others. Or the survival of an individual could require the simultaneous action of several others; for example, the reproductive success of the queen ant might require the coordinated action of groups of specialized workers. Experiments have found that the fitness effects of cooperative behaviors in microbes are not additive. It is clear that in general fitness effects cannot be assumed to be additive.

Two Approaches to Inclusive Fitness

Within the literature on inclusive fitness, there are two approaches for dealing with the limitation of additivity. The first approach is to restrict attention to simplified models in which additivity holds. For example, William D. Hamilton's original formulation of inclusive fitness theory includes additivity as an assumption. Additivity also follows from assuming that mutations have only small effects on phenotypes, and that fitness varies smoothly with phenotypes.

M. A. Nowak, C. E. Tarnita, and E. O. Wilson investigated the mathematical foundations of this first approach. They demonstrated that this approach also requires a number of restrictive assumptions beyond additivity of fitness effects, and is therefore applicable only to a limited subset of evolutionary processes. In response, more than one hundred authors signed the statement that "inclusive fitness is as general as the genetical theory of natural selection itself." How are we to understand this apparent contradiction?

The answer is that the above statement rests on a second,

alternative approach, which deals with the additivity problem in retrospect. In this approach, the outcome of natural selection must already be known or specified at the outset, and the objective is to find additive costs and benefits that would have yielded this outcome—regardless of whether they correspond to actual biological interactions. The cost (C) and benefit (B) are determined using the linear regression. The change in gene frequency is then rewritten in the form $BR - C$, with R quantifying relatedness. This regression method was introduced by Hamilton in a follow-up to his original work on inclusive fitness theory, and has been subsequently refined into a recipe for rewriting frequency changes in the form of Hamilton's rule.

The regression method underpins many claims of the power and generality of inclusive fitness theory. For example, it is often claimed that the regression method allows inclusive fitness to eschew the requirement of additivity. It is also claimed that the regression method generates a prediction of the direction of natural selection, and leads to a quantitative understanding of any frequency change as a consequence of social interactions between related partners.

Here we evaluate these claims by asking what, if anything, the regression method reveals about a given evolutionary change. We show that claims of the method's predictive and explanatory power are false, and the claim of its generality is not a meaningful one that could be evaluated. These findings call into question the idea that inclusive fitness provides a universal design principle for evolution—indeed, no such design principle exists.

Regression Method Does Not Yield Predictions

We now evaluate the various claims made regarding the regression method, starting with the claim that it predicts the direction of selection. This claim cannot be true, because the allele frequency change over the considered time interval is specified at the outset. The "prediction" merely recapitulates what is already known, such that the sign of $BR - C$ agrees with the predetermined outcome.

The regression method also does not predict what will happen over different time intervals or under different conditions. With any change in the considered scenario or time interval, the starting data must be respecified and the method reapplied, yielding new and independent results.

This lack of predictive power is unsurprising. It is logically impossible to predict the outcome of a process without making prior assumptions about its behavior. In the absence of any modeling assumptions, all that can be done is to rewrite the given data in a different form.

Experimentalists have noticed this absence of predictive capacity. One recent study applied the regression method to the cooperative production of an agent needed for antibiotic resistance in *Escherichia coli*. The authors conclude that "even if one has measured the values of B, C and R for a particular system of producers and nonproducers, one cannot predict what will result from changing either the structure of population or the biochemistry of the individuals."

Regression Method Does Not Yield Causal Explanations

We now evaluate the explanatory power of the regression method. The current literature appears to disagree on this point. Some works claim the method yields causal explanations for frequency change, whereas others make the more limited claim that it provides a useful conceptual aid. Moreover, the quantities that result from the regression method are commonly described in terms of social behaviors such as altruism and spite, imbuing these quantities with a "causal gloss" even if no direct claims of causality are made.

The claim that the regression method identifies the causes of allele frequency change cannot be correct, because regression can only identify correlation, and correlation does not imply causation. Moreover, because the regression method attempts to find additive social fitness effects that match given data, we should expect it to yield misleading results when social interactions are not additive, or when fitness variation is caused by other factors. Based on this principle, we present three hypothetical scenarios in which the regression method mischaracterizes the reasons for frequency change.

In the first hypothetical scenario, a "hanger-on" trait leads its bearers to seek out and interact with individuals of high fitness. We suppose that these interactions do not affect fitness. However, this seeking-out behavior leads fitness to become positively correlated with having a hanger-on as a partner; thus the regression method yields $B > 0$. According to the proposed interpretation, hangers-on should be understood as cooperative, bestowing high fitness on their partners. However, of course this gets causality backward—the high fitness causes the interaction, not the other way around.

Variants of this hanger-on behavior may occur in many biological systems. A bird may choose to join the nest of a high-fitness pair, with the goal of eventually inheriting the nest. Similarly, a social wasp may be more likely to stay at its parents' nest if the parent has high fitness, also with the goal of eventual inheritance. Applying the regression method to these situations would lead one to mistake purely self-interested behaviors for cooperation.

The second example is a "jealous" trait. Jealous individuals seek out high-fitness partners and attack them with the aim of reducing their fitness. We suppose that these attacks are costly to the attacker but only mildly effective, so that the attacked individuals still have above-average fitness after the attacks. The regression method yields B, $C > 0$, suggesting that the jealous individuals are engaged in costly cooperation. Again, this interpretation is wrong: the attacks are harmful, and the positive fitness correlation is due to the choice of interaction partners and the ineffectiveness of the attacks.

The third example is a "nurse" trait. A nurse will seek out

low-fitness individuals and make costly attempts to improve their fitness. We suppose, however, that this aid is only mildly effective, so that the aided individuals still have below-average fitness. The regression method yields $B < 0$, $C > 0$, misinterpreting this remaining low fitness as due to costly sabotage on the part of the nurses.

"Assumption-Free" Approaches

Finally, we turn to the claim that inclusive fitness theory is "as general as the genetical theory of natural selection itself." The argument is that, because the regression method can be applied to an arbitrary change in allele frequency (regardless of the actual causes of this change), it follows that every instance of natural selection is explained by inclusive fitness theory.

However, as we have seen, the regression method yields a "just-so story" that does not predict or explain anything about the given scenario or any other. Of course, there can exist cases for which the regression method yields correct causal explanations, and there can also exist cases for which the results obtained for one scenario are approximately accurate for certain others. However, the regression method provides no criteria to identify these cases—indeed, to formulate such criteria would require additional assumptions about the underlying processes. Without such assumptions, the results of the regression method do not answer any scientific question about the situation under study. The claim of generality is therefore meaningless.

This lack of utility is not due to any technical oversight. Rather, it arises from the attempt to extend Hamilton's rule to every instance of natural selection. This impulse is understandable, given the intuitive appeal of Hamilton's original formulation. However, the power of a theoretical framework is derived from its assumptions, thus a theory with no assumptions cannot predict or explain anything. As Wittgenstein argued in his *Tractatus Logico-Philosophicus*, any statement that is true in all situations contains no specific information about any particular situation.

There Is No Universal Design Principle

The concept of inclusive fitness arises when one attempts to explain the evolution of social behavior at the level of the individual. For example, inclusive fitness theory seeks to explain the existence of sterile ant workers in terms of the behaviors of the workers themselves. The proposed explanation is that workers maximize their inclusive fitness by helping the queen rather than producing their own offspring.

The claim that evolution maximizes inclusive fitness has been interpreted as a universal design principle for evolution. This claim is based on an argument by Hamilton that evolution maximizes the mean inclusive fitness of a population, and a separate argument by Alan Grafen that evolved organisms act as if to maximize their inclusive fitness. Both of these arguments depend on restrictive assumptions, including additivity of fitness effects. Because experiments have shown that fitness effects in real biological populations are nonadditive, these results cannot be expected to hold in general. Moreover, both theory and experiment have shown that frequency-dependent selection can lead to complex dynamic phenomena such as multiple and mixed equilibria, limit cycles, and chaotic attractors, ruling out the possibility of general maximands. Thus, evolution does not, in general, lead to the maximization of inclusive fitness or any other quantity.

Commonsense Approaches to Evolutionary Theory

Fortunately, no universal maximands or design principles are needed to understand the evolution of social behavior. Rather, we may rely on a straightforward genetic approach: Consider mutations that modify behavior. Under which conditions are these mutations favored (or disfavored) by natural selection? The target of selection is not the individual, but the allele or the genomic ensemble that affects behavior.

To investigate these questions theoretically, one needs modeling assumptions. These assumptions can be highly specific, applying only to particular biological situations, or broad, applying to a wide range of scenarios. Modeling frameworks that rely on general (yet precise) assumptions have recently emerged as a powerful tool for studying the evolution of populations structured spatially, by groups, and physiologically; the evolution of continuous traits; and inclusive fitness theory itself (in cases where fitness effects are additive and other requirements are satisfied). Although these frameworks can be used to obtain general results, none of them is universal or assumption-free. Instead, they draw upon their assumptions to make well-defined, testable predictions about the systems to which they apply.

Discussion

Inclusive fitness theory attempts to find a universal design principle for evolution that applies at the level of the individual. The result is an unobservable quantity that does not exist in general (if additivity is required) or has no predictive or explanatory value (if the regression method is used). If instead we take a genetic perspective and ask whether natural selection will favor or oppose alleles that modify social behavior, there is no need for inclusive fitness.

The dominance of inclusive fitness theory has held up progress in this area for many decades. It has consistently suppressed reasonable criticism and alternative approaches. In particular, the attempt to eschew the requirement of additivity using regression methods has led to logical obfuscation and false claims of universality. Reasonable inclusive fitness calculations that assume additivity represent an alternative method to account for fitness effects in some limited situations, but this method is never necessary and often needlessly complicated. There is no problem in evolutionary biology that requires an analysis based on inclusive fitness.

Having realized the limitations of inclusive fitness, sociobiology now has the possibility to move forward. We encourage the development of realistic models grounded in a firm understanding of natural history. With the aid of population genetics, evolutionary game theory, and new analytic procedures to be developed, a strong and resilient sociobiological theory can emerge.

Acknowledgments

I am grateful to John Taylor (Ike) Williams for his unwavering support and advice, to Robert Weil for his editorial guidance on this as on my earlier books published by W. W. Norton, and to Kathleen M. Horton for her invaluable assistance in research, editorial work, and manuscript preparation.

Chapter 2, "Solving the Riddle of the Human Species," is a modification of the author's "The Riddle of the Human Species," in *The New York Times Opinionator*, February 24, 2013. Chapter 3, "Evolution and Our Inner Conflict" was modified from an article of that name by the author, in *The New York Times Opinionator*, June 24, 2012. Chapter 11, "The Collapse of Biodiversity," is a modified version of "Beware the Age of Loneliness," in *The World in 2014, The Economist*, November 2013, p. 143.

Index